Applied plant anatomy

Applied plant anatomy

D. F. Cutler BSc, PhD, DIC

Head of the Plant Anatomy Section
Jodrell Laboratory
Royal Botanic Gardens
Kew

Longman
London and New York

Longman Group Limited London

Associated companies, branches and representatives
throughout the world

Published in the United States of America
by Longman Inc., New York

© Longman Group Limited 1978

First published 1978

Library of Congress Cataloging in Publication Data
Cutler, David Frederick, 1939–
  Applied plant anatomy.

  Includes bibliographies and index.
  1. Botany—Anatomy.   I. Title.
  QK641.C866   581.4   77-3290
  ISBN 0-582-44128-5

Printed in Great Britain by
William Clowes & Sons, Limited
London, Beccles and Colchester

Contents

Contents

Preface

Works on plant anatomy can be divided into two main categories; the advanced, largely theoretical text-books for university students and the reference books containing comparative, taxonomic and detailed descriptive information. Then there are the thousands of individual research papers, and the largely 'visual' books containing sets of photographs or drawings with a minimum of supporting text.

It is not surprising that the student beginning botanical studies comes to think of plant anatomy as being essentially an academic subject. In this book I hope to show some of the numerous ways in which plant anatomy can be applied to solve many important everyday problems. I have assumed a basic biological background for most readers, but have attempted to start at a straightforward level and develop the terminology and descriptions to the point where the significance of the applications can be understood. For those who need it, an illustrated glossary is put in a prominent place (Chapter 3) and not at the end of the book. This meant that a great deal of tedious definition could be left out of the main text.

I hope that in reading through this book the student will acquire a sound basis in terminology and an understanding of the main aspects of plant anatomy in a relatively painless way. References do not interrupt the general text, but useful advanced books and works of reference are listed at the end of each chapter.

Vegetative anatomy is stressed throughout. To have included extensive accounts of floral anatomy and embryology would have meant reducing the coverage of other aspects for which information is less readily available.

Since the student is encouraged to use this book to complement more general books on plant anatomy, it has been possible to get away from a number of the now classical plants which seem to serve universally as examples of particular characters. Examples have been drawn from an enormous collection of reference microscope slides held at Kew. The range of plants used should enable the student in the tropics to find examples familiar to him as easily as the person living in temperate regions. If the actual examples cited in the body of the text are not enough, please refer to the lists at the end of the appropriate chapters of fairly common plants from various parts of the world. I have picked out just a few features that you might expect to find in these examples. It would be a useful exercise to try to write your own, fuller anatomical descriptions of some of these plants.

The first chapter explains some of the methods and staining techniques that have passed the test of time. These are used routinely in the Jodrell Laboratory, Royal Botanic Gardens, Kew.

Dr C. R. Metcalfe, to whom I owe so much of my anatomical training, has read the manuscript and suggested many useful modifications and has most kindly allowed me to include an additional list compiled partly from his collected data for the revision of his famous volumes written with Dr L. Chalk, *Anatomy of the Dicotyledons.* This list shows a few selected anatomical characters and the families where they commonly occur.

I am indebted to Professor J. Heslop-Harrison, Director of the Royal Botanic Gardens, Kew, for the use of facilities at the Jodrell Laboratory out of 'official' hours to enable me to write this book.

From time to time I have referred to 'the laboratory' in the text; this means 'the Jodrell Laboratory'. This book was written independently of the Royal Botanic Gardens, Kew and the views which are expressed here do not necessarily reflect those of the Ministry of Agriculture, Fisheries and Food, or the Director, Royal Botanic Gardens, Kew.

Figures 4.7, 4.8 **B** and **C**, 4.18, 5.8, 7.5, 8.1 **A** and **B**, 8.2, 9.2 **A** and **B**, 9.3, 9.4, 9.6 and 9.7 are Crown Copyright, and are reproduced by permission of the Controller, Her Majesty's Stationary Office, and the Director, Royal Botanic Gardens, Kew.

I am indebted to Dr E. Ancibor for Fig. 4.33, Miss C. Brighton for Figs. 8.1 **A** and **B**, and Dr S. Owens for Fig. 8.2.

I am grateful to Mr L. Forman, Dr R. Harley, Mr F. N. Hepper, Dr A. Kanis, Mr A. Radcliff Smith and Mr C. C. Townsend who provided lists of species commonly available in various countries. The photographs were kindly processed in his own time by Mr T. Harwood, and Miss M. Gregory read parts of the manuscript. I should also like to thank the typists, Mrs S. Goddard and Miss M. Long.

Finally, I dedicate this book to my wife, Susan, for her patience and encouragement, which have helped me to complete it.

David F. Cutler
January 1977

Diagrams

Key to shading used in all diagrams throughout the book.

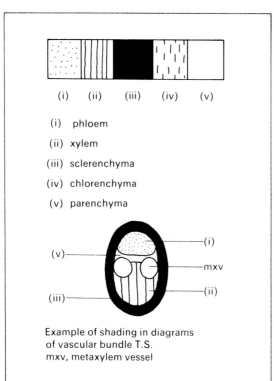

(i)   (ii)   (iii)   (iv)   (v)

(i)   phloem

(ii) xylem

(iii) sclerenchyma

(iv) chlorenchyma

(v)   parenchyma

Example of shading in diagrams
of vascular bundle T.S.
mxv, metaxylem vessel

Cover illustration

Clematis montana var. *rubens*, stem T.S. Crown Copyright. Reproduced by permission of the Controller, Her Majesty's Stationery Office, and the Director, Royal Botanic Gardens, Kew.

1 Introduction

The aim of this book is to present the fundamentals of plant anatomy in a way which emphasizes their application and relevance to modern botanical research. It is intended mainly for students at the intermediate level of a first degree course, but provides a readily understandable account of plant structure for the less advanced student.

A certain amount of jargon has to be learned in order to get to grips with any subject, and here I make no excuse for using terms which are specialized in their meaning. The correct use of technical terms aids clear thought and helps to make plant anatomy as exact as possible. Rather than define these words as they arise, and in so doing break the flow of thought, I have put those which are most used into an illustrated glossary which forms Chapter 3.

Examples are chosen from a wide range of plants from temperate to tropical environments – far too many textbooks neglect the rich tropical flora. Teachers should find plants mentioned which are readily available to them to illustrate particular cells or tissues. I hope those in tropical countries will seize the opportunity to look at plants growing on their own doorsteps, instead of having to send to north temperate lands for microscope slides of unfamiliar plants.

Applied anatomy is the key expression in this book. Plant anatomy is regarded as a dull subject by many students because the tradition has been to teach it as a catalogue of cell and tissue types with only slight reference to function and development, and no mention of the day-to-day use to which this knowledge is put in many laboratories round the world. Textbooks have been written to suit this more usual style of teaching. These advanced texts are of excellent value for the specialist student, but can be daunting to the relative beginner. Complementary to these books are those consisting largely of illustrations. These are of much benefit to students struggling to recognize what they see down the microscope, but again have the shortcomings that they mainly serve to teach a set of descriptive terms.

At Kew, plant anatomy is in everyday use as a tool to help in solving baffling problems – many of economic value and a good number of scientific interest. As such, the subject becomes alive and fascinating. We also apply anatomy to help solve rather more academic questions of the probable relationships between families, genera and species. The incorporation of anatomical data with the findings from studies on gross morphology, pollen, cytology, chemistry and similar disciplines enables those making revisions of the classification of plants to produce more natural systems. The economic significance of accurate classification and hence accurate identification of plants is frequently overlooked. The plant breeder, the food grower, the ecologist and conservationist all need accurate names for the subjects of their study. The chemists and pharmacognosists searching for new chemical substances must certainly know exactly which species or even which varieties yield valuable substances. Without an accurate name and description, they cannot repeat their experiments, or obtain further plant material of the same species, or know which closely related plants might be examined for similar properties. In this book, many more reasons indicating the need for accurate identification of plants will be given from everyday examples.

Materials and methods

Materials
Any readily available plant can be used to teach or learn plant anatomy. Over the years a small number of species have become 'approved' as standard plants for study. This has produced a stultifying effect. The plants chosen are thought of as 'typical', but often

they are quite atypical. Many botanists go through their lives thinking that maize is a typical mono-cotyledon – but grasses in general are very specialized and represent a very restricted view of monocotyledons as a whole. I am sure we frequently adhere to peas, lettuce, maize and sunflower in our physiological work because botanists are often unaware that other plants which can be grown with equal ease, are more varied and interesting in their structure.

Material is best if collected fresh. It can be examined fresh for cell contents, cytoplasmic movement and so forth. For ease of sectioning and histological studies it is better to 'fix' the fresh material by chemical means. Fixatives, when correctly formulated, will kill the plant material, preserve its shape and size, and render the tissues suitable for sectioning.

Fixatives

Seventy per cent alcohol is typically used as a plant fixative in schools, since it has little effect on the student should he pour it over himself. This reagent hardens the plant tissues and can cause changes in shape. The zoologist soon learns to live with formaldehyde, which has to be treated with more respect. There is no reason why students should not use more potent liquids for fixing plants if suitable care is taken. For general histological purposes, the following mixture is found to be excellent:

Formalin–acetic–alcohol or FAA
850 ml   70 per cent alcohol
100 ml   40 per cent formaldehyde
50 ml   glacial acetic acid

This is a corrosive liquid, and should it come into contact with the skin it should be washed off immediately. It is well worth the trouble and care to use material fixed in FAA because it sections well and can be kept in the reagent indefinitely. **The fumes are harmful and should not be inhaled.**

Material to be fixed is normally cut into portions to enable rapid penetration of the fixative. The portions should be of such a size that they can be readily identified and oriented. Bottles with wide mouths and plastic screw tops are ideal for storage, and can be obtained in a range of sizes. It is best to keep the plant in fixative for at least 72 hours before use. Plant material in the FAA can be stored for as long as required, but the bottles should be inspected regularly for evaporation and topped up with 70 per cent alcohol if necessary. This is the most volatile of the constituents.

Specimens to be sectioned are removed with forceps and washed in running tapwater for $\frac{1}{2}$ to 1 hour. They can then be handled safely.

Dried herbarium material can often be used for anatomical studies. Some plants revive easily, but others are unsatisfactory. If there is no fresh material available, then dried material can be fixed after boiling in water for 5–15 minutes and cooling. Some people add a few drops of detergent to aid wetting.

Sectioning

The safety razor blade can be used to produce sections thin enough for study under magnifications of 100 to 400 ×, or sometimes more. Practice is needed. Those with a steady hand will get better sections with the thinner, double-sided razor blade, but very thin blades are too flexible. Every anatomist bears the scars of early battles with tough plant material.

For the production of large quantities of slides of the same specimen for class study, a more refined method of sectioning is needed. A rotary or rocking microtome is often used, but the method which involves embedding the specimen in a block of wax or similar material is almost entirely unnecessary for histological studies. It is only indispensable for sections under 10 μm thick, or for subjects such as flower buds where the various parts would separate and become disarranged when not embedded for sectioning.

Normally, sections of between 15–30 μm are suitable for histological study. These can be cut with a sledge or sliding microtome. The Reichert OME is such an instrument. It is highly recommended. It is one of the types in which the specimen to be sectioned is firmly held in a universal clamp (to allow for correct orientation) and the knife is brought towards and over the specimen. Models in which the knife is fixed, and the specimen is made to move past the knife are not so universally useful.

The moving knife sledge microtome can be used for the following types of material, using 50 per cent alcohol for lubrication (applied to the knife blade with a camel-hair brush).

1. Tough materials, such as wood, are cut into 1 cm cubes, with an orientation as described on p. 8. The cubes are boiled in water until waterlogged, i.e. they sink when cold water is added to the container. The cubes are removed, cooled and clamped. If they are very hard, they may need to have a jet of steam directed on the surface to be cut (see Fig. 1.1), but often they are soft enough to cut directly. Some woods contain silica, a substance so hard that it rapidly blunts the microtome knife. Silica can be removed by standing the wood for 12 or more hours in 10 per cent hydrofluoric acid, in plastic containers. **Very great caution should be taken with this acid. It causes**

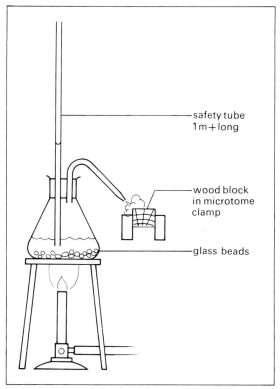

Fig. 1.1 Apparatus for producing a steam jet to soften wood prior to sectioning.

cut across the diameter at an angle (Fig. 1.2E–H). When the two parts are placed side by side with the material to be sectioned at the bottom of the V, clamping causes the outer parts of the cork to roll outwards, but the material is retained. If the disc was cut to produce a flat surface before clamping, the outward curving would release the specimen, as in Fig. 1.3.

3. Leaves to be sectioned transversely are rarely just the right width for the clamp. Wider leaves can be folded once or several times so that they form a sandwich in the cork, Fig. 1.4. With narrow leaves, it is best to put several leaves between the cork slices – there is more chance of getting some good sections.

Most mesophyte leaves section easily. Some succulent or very soft hydrophyte leaves and stems can pose problems. In these plants the cells are thin-walled and burst easily if compressed when turgid. A simple and usually effective remedy is to allow the specimen to go limp on the bench for half an hour or so. It can then be firmly clamped, cut, and the sections put in water. If they do not return to their natural shape in water, 50 per cent alcohol may be used, or even, for a second or so, immersion in un-diluted bleach such as 'Domestos' or 'Parozone' (sodium hypochlorite) will cause them to return to the uncompressed form.

Certain leaves contain silica bodies (particularly those of grasses and sedges) which blunt the micro-tome knife and tear the section. So if a section appears torn, first examine it to see if silica bodies are present. Hydrofluoric acid (10%) can be used to remove the silica bodies, **but must be treated with the utmost caution.**

As each section is cut it will slide onto the knife blade, lubricated by the 50 per cent alcohol. It should be lightly transported on a paint brush to a petri dish of 50 per cent alcohol. The section can be examined temporarily in water, or will keep for several months in 50 per cent glycerine solution on a slide (stored flat). Starch distribution can be studied in such sections, and chloroplasts and other larger cyto-plasmic inclusions can be seen.

Clearing

Sometimes it is an advantage not to have cell contents obscuring the tissue distribution but, before 'clearing' the sections, some should be studied with their inclusions. Sections can be cleared by trans-ferring them from the 50 per cent alcohol to a dish of water with a brush or fine forceps. Then, using a mounted needle or fine forceps they are placed in a cavity block containing undiluted Parozone, Domes-tos or other sodium hypochlorite household bleach.

serious burns even in low concentration, and the burns heal very slowly. It is not recom-mended for class use, but could be used by trained technicians. After treatment, the wood must be washed in running water for several hours.

2. Twigs may need to be boiled before sectioning. Softer stems are best fixed and washed before sectioning. Many cylindrical objects need some sup-port in the microtome clamp. Support is normally provided in the form of cork or pith. Pith tends to be-come soggy when wet, and cork can contain unex-pected sclereids which blunt the knife, but on the whole I prefer cork. Suitable bottle corks with few lenticels should be selected (Fig. 1.2A–D). Circular slices about 3–4 mm thick are cut off, using a razor blade. The discs of cork are cut across the diameter and the two halves placed side by side.

For transverse sections of stems, a notch cut in one half of the cork disc will help to keep the specimen correctly oriented, without allowing it to be com-pressed excessively. Alternatively, the cork may be cut along its long axis, and the two halves used to mount the specimen to be sectioned.

For making longitudinal sections, the cork slice is

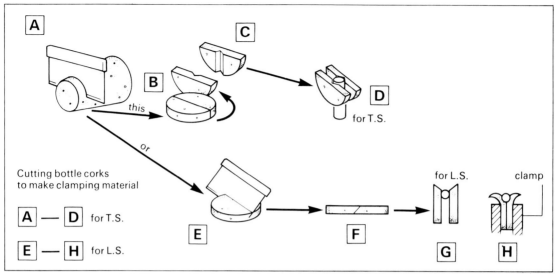

Fig. 1.2 Preparing a cork for holding material to be sectioned. A–D for T.S., E–H for L.S.; note that the oblique cut in cork E helps to prevent cylindrical stems from being released from the cork on clamping.

The time needed for cell contents to dissolve away varies from subject to subject and can be determined by visual inspection. It usually takes about 5 minutes. The whole section will dissolve if left long enough! After immersion in the bleach, the sections are thoroughly washed in water. Take care not to get the brush in the bleach – its bristles will dissolve.

Staining

After thorough washing, the sections are ready for staining. Two main types of stain can be used, (*a*) those which are temporary, whose colour fades, or which gradually damage the section and (*b*) those which are regarded as permanent. Even permanent stains may lose their colour if exposed to sunlight, so slides should be stored in the dark.

With care, stains can be selected to give the maximum contrast between the various cell and tissue types in the plant. They might be selected because they colour particular parts of the cell wall structure and indicate its chemical composition. The stains described below are in daily use in the Jodrell Laboratory. Those who want comprehensive lists should see the books by Gurr, Foster or Peacock – to mention but three of the many guides to microtechnique.

1. *Temporary stains*

(*a*) 1 per cent aqueous methylene blue All cell walls turn blue, except cutin or cutinized walls which remain unstained; cell walls take up a degree of intensity of blue depending upon their chemical composition and physical structure; various wall layers frequently stain differently.

The stain may be mixed with 50 per cent glycerine, about 10 per cent of 1 per cent aqueous stain to 90 per cent of 50 per cent glycerine, and sections mounted directly into this medium. This mixture is also useful for staining macerated tissues which are difficult to handle. A drop of washed macerate in water is mixed with a drop of the mixture on a slide, and the cover slip put on.

(*b*) Chlor–zinc–iodine solution (Schulte's solution) This solution consists of: zinc chloride 30 g, potassium iodide 5 g, iodine 1 g and distilled water 140 ml. Cellulose walls turn blue, starch turns blue-black, lignin and suberin turn yellow and moderately lignified walls turn greeny blue.

Sections are placed on the slide, and a drop or two of CZI added. This can be drawn off and replaced by 50 per cent glycerine after 2–4 minutes, but satisfactory results can be obtained by adding 50 per cent glycerine directly, and mounting in the mixture.

This stain swells the walls and eventually dissolves them. Consequently, care must be taken when describing wall thickness.

(*c*) Chlorazol black – saturated solution in 70 per cent alcohol Stains walls black or grey; particularly good for showing pitting.

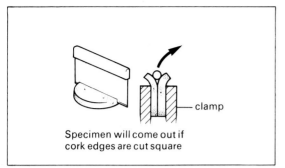

Fig. 1.3 The wrong way to cut cork for making L.S. of material. When clamped, the cork curls back and the specimen is released.

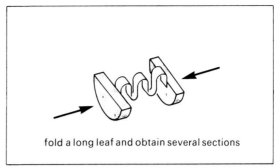

Fig. 1.4 Long leaves can be folded several times before sectioning. Several sections will then be obtained with each cut.

(*d*) Saturated carbolic acid solution Sections are mounted directly in the solution (**which should be kept off the hands**). Silica bodies usually turn pink; this helps to distinguish them from crystals which remain colourless.

(*e*) Phloroglucin and conc. HCl The phloroglucin is added to the section, and then the HCl. Lignin turns red.

(*f*) Sudan IV Sections can be mounted directly in the stain. Fats, cuticle turn orange.

(*g*) Ruthenium red Mucilage and some gums turn pink. Sections can be mounted directly in the stain.

2. *Permanent stains*

Safranin (1 per cent in 50 per cent alcohol) and Delafield's haematoxylin. Cellulose turns dark blue, lignin turns red and cellulose walls with some lignin turn purple.

Freshly mix the safranin with matured Delafield's haematoxylin in the proportions of 1:4; filter. The stock mixture can be used for up to about one week, but should be filtered before use each day.

Sections should be transferred from water (after washing all bleach away) into a cavity block containing the stain, and the block covered with a glass lid. Most sections take up the stain in 2–6 hours; some need less time. They are then transferred to a petri dish of 50 per cent alcohol containing 2–3 drops of conc. HCl. This solution removes the stain, acting on the safranin first. The object is to obtain a satisfactory colour balance, and only experience will tell when this has been reached.

In general, sections should be removed when they still appear to be slightly dark or overstained for best

results, the colours look less intense under the microscope. The decolorizing action is halted by placing the sections into a petri dish of 95 per cent alcohol. After about 5 minutes they can be transferred to absolute alcohol in a covered petri dish. Five minutes later they can be transferred either to a 50/50 mixture of absolute alcohol/xylene in a covered dish, or this step may be eliminated and they may be transferred directly into xylene. **Xylene fumes should not be inhaled.** After ten minutes in the xylene, the sections can be mounted in Canada balsam on the microscope slide. Any milkiness in the section at this stage means that water is still present, and the section should be taken back through xylene, then fresh absolute alcohol, fresh absolute alcohol/xylene and fresh xylene before re-mounting. Sections which curl up or roll up should be straightened out in 50 per cent alcohol. As they progressively dehydrate in purer alcohol they become more brittle and cannot be unrolled without breaking. Curled wood sections can be flattened by drawing them over the edge of a slide partly immersed in 50 per cent alcohol (Fig. 1.5) – a process needing three hands! Alternatively a section lifter can be used. Once on the slide they can be 'set' using a few drops of 95 per cent alcohol.

If it is more convenient to stain overnight, the safranin/haematoxylin mixture can be used in the proportions of 94:6. Although fast green can be used as a counter-stain for safranin, we have found that haematoxylin produces a colour which photographs better on normal panchromatic film.

Fast green can be used on its own as a stain for macerated material. The macerate is dehydrated by decanting alcohols of 50, 70, 90, 99 per cent and absolute in turn from a tube containing the macerate. It is a help if a small hand centrifuge can be used to settle the cells at each stage. Finally, the cells are transferred to a slide bearing 2–3 drops of euparal containing 2–3 drops of fast green per 10 ml. The cover slip is applied.

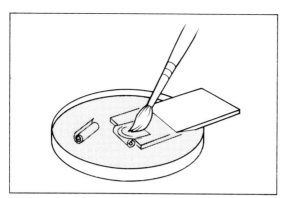

Fig. 1.5   Drawing a curled section onto a microscope slide.

Mounting media

Neutral Canada balsam has the advantage that it is not likely to remove safranin from the sections. Some modern substitutes, whilst being less yellow, are too acid and gradually cause the safranin to leach out, or contract markedly if over-dried.

Euparal is used either where it is undesirable to pass very delicate sections through xylene after absolute alcohol because they may distort, or for macerated material, as described above.

Slides are baked flat, in an oven at 58°C for 10–14 days, to thoroughly dry the mounting media. Those sections in Canada balsam are firmly embedded at this stage, and the slides can be stored upright. Those in Euparal, however, may be set round the edges only, and should be treated with care, or given further baking.

For upright storage of slides, aluminium holders are very convenient. Four slides are housed in each holder. The holder is the size of an index card and can be kept in a standard card filing cabinet.

Preparation of surfaces

The staining methods applied to sections can equally well be used for surface preparations.

Leaf surfaces

The epidermis of most leaves can be readily removed by the scraping method. Only those leaves with very prominent veins, or large, numerous hairs pose problems, and demand a great deal of patience.

Material may be fresh, or washed after fixing. A suitable piece is cut from the leaf (Fig. 1.6). The surface for study is placed face down on a glazed tile or glass plate. It is irrigated with a few drops of sodium hypochlorite. One end is held securely with a cork, and the other end scraped lightly with a safety razor blade. With practice a double-edged blade can be used, but it is better to start off by using a single-edged blade. The blade is held at 90° to the leaf (Fig. 1.6). The gentle scraping is continued, adding extra hypochlorite as necessary, keeping the leaf well irrigated. If the leaf is not severed by a forceful scrape, you will end up with a thin, clear area which can be cut off, placed in a cavity block for a few minutes with sodium hypochlorite and then washed in a petri dish of water. Loosely adhering cells can then be brushed off with a camel-hair paint brush.

The preparation can then be viewed in water under the microscope. It is easy to see if enough has been scraped away. Experience will soon enable you to judge when the end point is reached in scraping. Make sure the surface is placed the right way up before the final mounting.

With some material it is unnecessary to scrape the leaf, since the epidermis can be removed by peeling it off the fresh leaf. This is done by folding the leaf to break the surface, and either stripping directly, pulling one part of the leaf downwards relative to the other, or by holding as thin a layer of the surface as possible between forceps and peeling it back.

Stem surfaces

A thin strip of stem surface can be obtained by making the first L.S. cut on the microtome just pass through the surface layers. This requires careful microtome adjustment, but is quite satisfactory.

Surface replicas

Sometimes it is not possible or desirable to remove the epidermis itself – a plant may be rare, or there may be little material readily available. A good imprint of the surface can be obtained with a film of cellulose acetate. It may be necessary to wipe the leaf surface with acetone to clean it. Then clear nail varnish is brushed on. More than one layer may be required. The dried film can be removed and mounted, preferably in a medium with a different refractive index or little will be seen.

Although replicas can be made easily, using a variety of materials (latex, for example), I prefer to work with the epidermis itself. Replicas can tell us little about cell contents.

Cuticular preparations

Various chemical treatments can be used to cause the cuticle to separate from the leaf. The cuticle is in many respects better than the replica, but can be very delicate.

One method is to digest away the leaf tissues using nitric acid (**care**!). The cuticles will frequently float

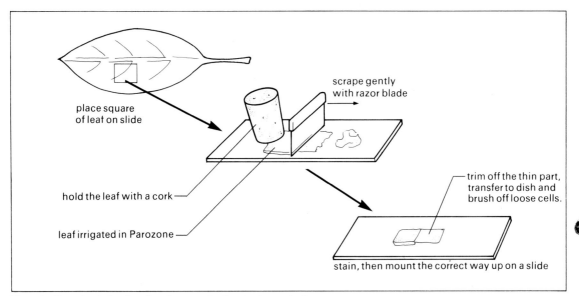

Fig. 1.6 Preparing leaf surface for microscopy by the scraping method.

to the surface, or, when the leaf has been fully washed, can be teased off the dissolving tissues.

Clearing material
Whole thin leaves or stems or flowers can be made transparent by soaking them in chloral hydrate, washing and soaking in sodium hydroxide solution, alternately, for several changes, with several hours at each stage.

After the final washing, the organ can be carefully stained in safranin (1% in 50% alcohol) dehydrated through a very gradual alcohol series and mounted. Veins and sclereids show up well.

Alternatively, the preparation may be left unstained for examination by various optical methods (p. 9).

Standard levels (Fig. 1.7)
When plants are being examined for comparative purposes, either for identification or for taxonomic reasons, it is important that similar parts of the organs are looked at. The leaf, for example, is normally looked at in transverse section across its broadest region, or half way along the length of the lamina. The surface of the leaf is studied near the central region of the lamina. The margin may also be examined.

Petioles should be examined in T.S. just where the lamina begins, half-way down its length and also near the base. Stems are normally sectioned in the middle of the internode, or, in addition, at the node. Roots

are normally sectioned at a convenient level, since accurately defined positions are harder to delimit. For very detailed studies, of course, sections are required from many other levels, and sometimes serial sections are needed. These are particularly useful in the study of nodes, and shoot apices.

A simple infiltration technique
Fix in FAA for 48 hours; wash in running water for 1 hour. Pass through alcohol series 50, 70, 90 and 95 per cent and absolute 4 hours each; absolute alcohol–xylol 3:1, 2 hours; absolute alcohol–xylol 1:1, 2 hours; absolute alcohol–xylol 1:3, 2 hours. Xylol, two changes, 1 hour each. Xylol:52° wax 3:1, 1:1, 1:3, pure wax two changes, 2 hours each, then overnight in pure wax, followed by one change for 2 hours in pure wax. The normal embedding procedure in the embedding oven is then followed.

Electron microscopes
This book is not intended for people who are using electron microscopes, but some space must be given to a brief description of their uses.

There are two main types of electron microscope, the transmission microscope (TEM) and the scanning electron microscope (SEM). Thin sections are examined in the transmission microscope, or carbon replicas produced from specimens can be used where surface features are to be studied. Electrons are made to pass like a focussed light beam through the section. Some parts of the specimen are electron-dense or are prepared by stains and fixatives to be electron-dense,

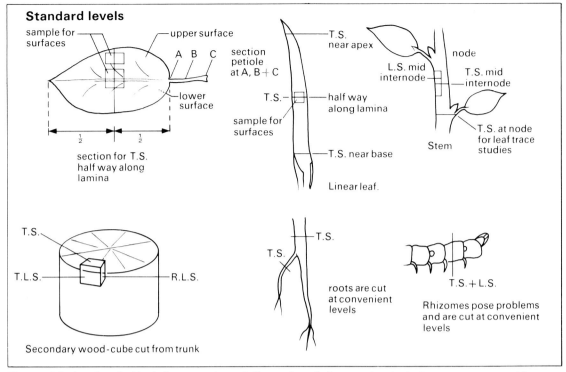

Fig. 1.7   Selection of standard levels for comparative work. For wood, a cube is prepared so that it gives transverse, tangential longitudinal and radial longitudinal faces.

whereas other parts are electron-opaque, permitting electrons to pass and form an image either on a special fluorescent screen, or directly on a photographic plate. The transmission microscope uses electromagnetic lenses to focus the electron beam, which has a very much higher resolving power than a beam of light. That is, it can make distinct points which are very close together on the object. The largest commonly available microscopes can resolve between points about 20 nm apart, and have the ability to magnify above 500,000 ×. Naturally, at such high magnifications, only very small areas can be seen at a time. By comparison, the best light microscope using green light can give a maximum real magnification of about 1,200 ×. Using ultraviolet light, slightly higher magnifications can be obtained.

Methods of preparation of the specimen for transmission microscopes, are straightforward, but a very skilled operator is needed to produce good results. Fixation of the specimen is very critical, for example.

The scanning electron microscope is used most commonly to examine the surface of specimens. Some specimens can be examined fresh for a brief time, but most are dehydrated carefully to minimize shrinkage and distortion, then coated with a very thin layer of metal, usually gold or gold/palladium alloy. This gives a better image and prevents contamination of the microscope by water.

Because of its relative ease of use, and since quite low (10 ×) magnifications as well as those of up to or above 180,000 × can be obtained on it, the instrument has a very wide use in applied plant anatomy. Currently, resolution of 1,000 nm or even better than 700 nm can be obtained as a matter of routine.

The specimen is bombarded with a focussed ray of electrons. The electrons are made to scan in parallel lines over a rectangular area. Secondary electrons are emitted by the object, are collected by a series of electronic devices, and a synchronous image is displayed on a small cathode ray tube. Most tubes are about 10 cm square, and have about 1,000 lines. The screen itself is photographed to produce a permanent record.

Provided the coated specimen is kept clean and dry, it can often be used many times.

A great depth of field can be obtained with this intrument, about 500 times that of a light microscope. Many surface patterns of leaves, seeds and fruits, spores, etc., are being seen and understood properly for the first time.

Of course, the cost will make it impossible for many people to own or even use an SEM, but once characters have been seen and illustrated or described, it is astonishing how many can be seen with a good dark field, epi-illuminating optical microscope. It is only specimens which have to be magnified above about 1,200 times that cannot be interpreted with the light microscope.

Extending the use of the student's microscope
It is unusual for the student to have a high quality light microscope, but the one available will serve well if it is kept clean, and all the lenses are centred properly.

One simple way to make the microscope more versatile is to make polaroid attachments. A disc of polaroid material mounted over the eyepiece, and another fitted in the filter carrier, or in a holder between the mirror (or light source) and the microscope slide, turn the instrument into a polarizing microscope. Crystals become easily observable, as do starch grains and details of cell wall structure. Figure 7.5 shows part of a stem section in polarized light.

Thin cellophane placed in one layer over the lower polaroid sheet (the analyser) will cause the light beam to become elliptically polarized. This phenomenon gives a coloured background against which crystals, etc. will appear a different colour. Rotation of the polaroid over the eyepiece (polarizer) will cause changes in the colours. This technique is useful for examining unstained, macerated material (p. 50), sclereids in cleared material (p. 41) or for looking at hairs or surface details in preparations where staining would make the subject too dense.

Other optical techniques
Other optical techniques include phase contrast, anoptral contrast (Fig. 4.7) and dark ground. Fluorescence microscopy and interference microscopy are also used in research.

The use of very thin (ultra microtome) sections has recently enabled the light microscopist to see details of cell walls otherwise obscure, and the study of transfer cells (p. 13) is possible with such sections.

Further reading

Bradbury, S., 1973. *Peacock's Elementary Microtechnique* (4th edn rev.), Edward Arnold, London.

Foster, A. S., 1950. *Practical Plant Anatomy*, Van Nostrand, New York and London.

Gurr, E., 1965. *The Rational Use of Dyes in Biology*, Leonard Hill, London.

Purvis, M. J., Collier, D. C., & Walls, D., 1964. *Laboratory Techniques in Botany*, Butterworths, London.

2 Basic morphology and tissue systems

Since each organ of the plant will be discussed in detail in later chapters, this section is intended only to remind the student of the basic arrangements of tissue systems. It is not intended to be comprehensive, and by its very nature it oversimplifies the complex and wide range of form and organization existing in the higher plants. Some terms are used without explanation. The glossary, forming an essential part of the book, should be consulted if the meaning of a term is unclear.

The main problems presented to a terrestrial plant are:

1. Mechanical, i.e. supporting itself in one way or another so that it can expose a suitable surface area with cells containing chloroplasts to the sunlight to intercept and fix solar energy.

2. The movement of water and minerals from the soil, via the roots to regions where they can be combined with other materials to build the plant body, and the movement of synthesized food material from the site of synthesis to places of growth or storage and from the stores to growing cells at the appropriate time.

3. Reproduction; placement of reproductive organs where the pollen or gamete receptor/mechanism can operate successfully, and after fertilization and spore/ seed production, ensure dispersal of the propagules.

4. Secondary growth in thickness.

The first two of the problems outlined above are dealt with by well-organized, if complex, systems in the higher plants, and will be summarized here in diagrammatic form. The third, reproduction, is outside the scope of this book and the fourth, secondary growth, is discussed in Chapter 5.

Mechanical support systems

Figure 2.1 illustrates the basic systems in a monocotyledonous plant, and Fig. 2.2 those in a dicotyledonous plant showing some secondary growth in thickness.

The systems can be divided into two main classes:

(*a*) The inflated or turgid, thin-walled cells; these are present in growing points, and the cortex and parenchymatous pith of many plants. They constitute the bulk of many succulent plants, e.g. *Aloe*, *Gasteria* leaves, *Salicornia* from salt marshes and *Lithops* from desert regions.

The cell wall acts as a very slightly elastic container; internal liquid pressure inflates the cell so that it becomes supporting, like the air in an inflated car tyre. Its support properties depend on water pressure, so a water shortage can lead to a loss of support and wilting. Some fairly large organs can be supported by this system, but they usually rely on the additional help of devices which reduce water loss, such as a thick cuticle, and perhaps also thick outer walls to the epidermal cells, and specially-modified stomata. A strong epidermis is particularly important, since it acts as the outermost boundary between the plant cells and the air. A split in the skin of a tomato, for example, rapidly leads to deformation of the fruit, or a cut in the succulent leaf of a *Crassula* or *Senecio* rapidly opens up.

Not many plants rely on the turgid cell/strong epidermis principle alone.

(*b*) Both monocotyledons and dicotyledons have specially-developed, elongated, thick-walled fibres, in suitable places, which assist in mechanical support. Alternatively, they may have specially thick-walled parenchyma (prosenchyma) or, in those primary parts of the stem where growth in length is continuing, collenchyma cells. Although there are only a few

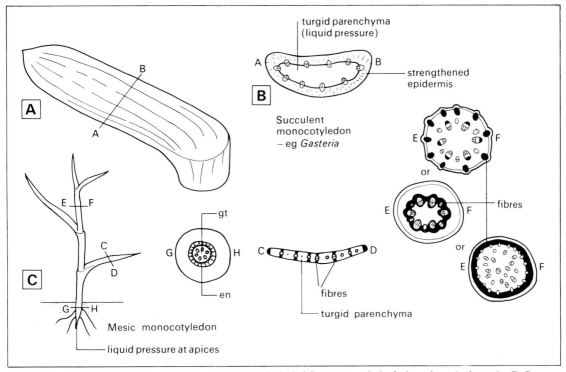

Fig. 2.1 Some mechanical systems in monocotyledons. A fleshy leaf of *Gasteria*; note lack of sclerenchyma in the section B. C a mesic monocotyledon, C–D shows one type of sclerenchyma arrangement in leaf T.S.; E–F show three of the main types of sclerenchyma arrangements in the stem T.S.; G–H shows a typical root section in which most strength is concentrated in the centre. en, endodermis; gt, ground tissue, which may be lignified.

common ways in which special mechanical cells are arranged in the stem, leaf or root, it is the variations on these themes which are of particular interest to those who have to identify small fragments of plants, or make comparative, taxonomic studies. The variations will be dealt with in detail in the chapters dealing with each organ. Obviously, to be effective the mechanical system must be economical in materials, and the cells must not be arranged in such a way as to hinder or impede the essential physiological functions of the organs.

The mechanical systems develop with the growth of the seedling. At first, turgid cells are the only means of support, but collenchyma may rapidly become established, particularly in dicotyledonous plants. As shown in Fig. 2.2 this tissue is concentrated in the outer part of the cortex, and is frequently associated with the midrib of the leaf blade, and the petiole.

Collenchyma is essentially the strengthening tissue of primary organs, or those undergoing their phase of growth in length. The cells making up this tissue have thickened cellulosic walls and are often found with chloroplasts in their living protoplasts.

Sometimes the only other mechanical support is provided by the wood (xylem) tracheids of the vascular system, as in most gymnosperms, or by the tracheids, vessels and xylem fibres of the angiosperms. However, far more commonly there are fibres outside the xylem (extraxylary fibres) which are arranged in strands or a complete cylinder, and can give considerable strength to herbaceous, and particularly herbaceous monocotyledon stems and leaves. The much elongated fibres, with their cellulose and lignin walls are not so flexible and do not stretch as readily as the collenchyma cells; consequently they are often found most fully developed in those parts of organs that have ceased growth in length.

Figure 2.1 shows some fibre arrangements in monocotyledon stems and leaves. In the leaf, fibres commonly strengthen the margins (e.g. *Agave*) and are found as girders or caps associated with the vascular bundles. In the stem, strands next to the epidermis can act rather like the iron or steel reinforcing rods in reinforced concrete. Together with a ribbed outline which they often confer on the stem section, they produce a rigid yet flexible system with economy of use of strengthening material.

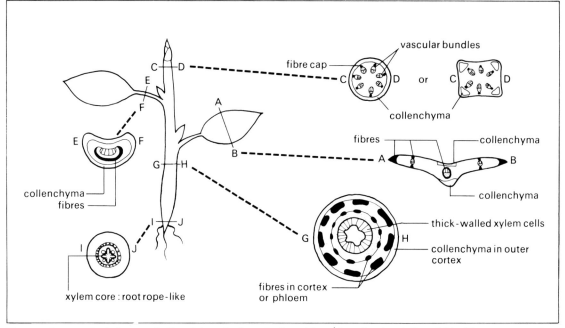

Fig. 2.2 Some mechanical systems in dicotyledons. A schematic plant with position of sections indicated. Liquid pressure occurs in turgid cells through the plant. Collenchyma is often conspicuous in actively extending regions and petioles. Sclerenchyma fibres are most abundant in parts which have ceased main extension growth. Xylem elements with thick walls have some mechanical function in young plants and give a great deal of support in most secondarily thickened plants.

Tubes are known to resist bending more effectively than solid rods of similar diameter; they also use much less material than the solid rod. It is not surprising then, that tubes or cylinders of fibres commonly occur in plant stems. They may be next to the surface, further into the cortex, or may occur as a few layers of cells uniting an outer ring of vascular bundles (Fig. 2.1).

The various arrangements will be discussed in more detail in Chapter 4, but mention must be made that in some monocotyledon stems individual vascular bundles, scattered throughout the stem, can each be enclosed in a strong cylinder of fibres. Each bundle plus its fibre sheath then acts as a reinforcing rod set in a matrix of parenchymatous cells.

Fibres or sclereids in dicotyledon leaves are also often related to the arrangement of the veins in the lamina, and also to the petiole vascular traces. These are shown in Fig. 2.2. The concentration of strength in a more or less centrally placed cylinder or strand in the petiole permits considerable torsion or twisting to take place as the leaf blade is moved by the wind, without damage occurring to the delicate conducting tissues. Primary dicotyledous stems may have fibres in the cortex and phloem.

The subterranean roots of both monocotyledons and dicotyledons have to resist different forces and stresses from those imposed on the aerial stems. These are tensions or pulling forces, as opposed to bending forces. The concentration of strengthening cells near the root centre gives it rope-like properties.

The transport systems

It is not possible to present a simple, comprehensive model to demonstrate the wide range of arrangements of vascular systems which occur in either dicotyledon or monocotyledons. Dicotyledons in which tissues are wholly primary tend to be a little more stereotyped than monocotyledons, but even then there is a very wide range of arrangements.

The essential elements of both systems are the xylem, concerned with transport of water and dissolved salts, and the phloem, which translocates synthesized but soluble materials around the plant, to places of active growth or regions of storage.

In the apex of the shoot and root, vascular tissue is not developed; soluble materials and water move from cell to cell in these unspecialized zones. Not far back from these growing points, however, more formal conducting systems are needed. Procambial strands, the precursors of the vascular bundles, are first seen and, then, further from the tips, phloem alone followed by phloem and xylem together. The newly formed strands in most dicotyledons join the previously formed vascular bundles through a leaf or branch gap.

In most dicotyledons the leaf lamina has a mid-rib to which are connected the lateral veins. The latter form a network with major and minor systems. The midrib is directly connected to the petiole trace. This enters the stem and joins on to the main stem system through a leaf trace gap, as described above. In the primary stem all the vascular bundles are separate from one another (they remain separate in many climbers, e.g. *Cucurbita, Ecballium*), but in most dicotyledons the bundles become joined into a cylinder by growth of xylem and phloem from a special cambium.

Where the systems of the stem and root meet a complex re-arrangement of tissues takes place in the primary plant. In the stem vascular bundles, the phloem is to the outer side of the xylem in the majority of plants. In the root, the xylem is central, and may have several lobes or poles; the phloem is situated between these. The transition region is called the hypocotyl. After secondary growth has taken place, this complex zone becomes surrounded by secondary xylem and phloem, and the shoot and root anatomy become more similar. Secondary growth is discussed in Chapter 5.

Transfer cells are specialized parenchymatous cells found in various parts of the plant in regions where there is a physiological need for transport of materials, but where the normal phloem or xylem cells are not in evidence. A good example is the junction between cotyledons and the shoot axis in seedlings. Transfer cells may also be present near the extremities of veins, or near to adventitious buds, for example. Thin sections of the walls of transfer cells show them to have numerous small projections directed towards the cell lumen. These greatly increase the possible area of cell wall–protoplasm interface, a site of metabolic activity concerned with the movement of materials from one cell to the next. The projections are so fine that conventional sections with a rotary microtome are too thick for them to be seen.

Monocotyledons are quite different from dicotyledons in their vasculature. Leaf and stem are commonly much less readily separable as distinct organs, and jointly constitute the shoot. There is no secondary growth by a true cambium, so a cylinder of vascular tissue cannot be formed. When secondary growth occurs, as in *Agave, Cordyline*, etc., it is by means of a special tissue, situated near to the stem surface, which

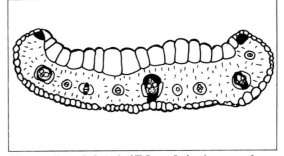

Fig. 2.3 *Juncus bufonius* leaf T.S., × 48, showing 1 row of vascular bundles, with the xylem poles directed towards the adaxial surface. Note the marginal sclerenchyma strands and the difference in size between adaxial and abaxial epidermal cells. Each small vascular bundle has a parenchyma sheath; in larger bundles sclerenchyma caps interrupt the parenchyma sheath.

forms complete, individual vascular strands and additional ground tissue.

Vascular bundles are usually arranged in the stem with the xylem pole facing towards the stem centre (but this is not invariably so). The arrangement of leaf vascular bundles is very variable; grasses and *Juncus* species, for example, often have one row as in Fig. 2.3. Some of the other types of arrangement are discussed in Chapter 4.

Since there is no vascular cylinder, where leaf traces (bundles) enter the stem they do not form gaps. They may join at nodes, where all the bundles at that particular level of the stem form a sort of plexus, e.g. *Aloes*. Sometimes, in stems with nodes, the leaf traces may continue downwards from their points of entry into the stem for a complete internode before joining the nodal plexus below, (e.g. *Restio, Leptocarpus*, Restionaceae). In other plants without nodes, e.g. Palms, the leaf traces follow a simple path curving inwards towards the stem centre, and then gradually 'move' towards the outer region of the stem lower down. These leaf traces join onto the main bundles by small, inconspicuous bridging bundles. This system is beautiful in its simplicity, but fearfully difficult to analyse because there are so many (several hundred) vascular bundles even in the narrow portion of a stem of a small palm like *Rhapis*. As one follows the course of palm bundles, they are seen to spiral down the stem.

The main root does not often develop in monocotyledons. Its function is usually taken over by numerous adventitious roots which arise at an early stage, and join the stem vascular system in what frequently appears as a jumble of vascular tissue with very short elements both in the phloem and xylem.

Further reading

General advanced texts
Cutter, E. G., 1969. *Plant Anatomy: Experiment and Interpretation.* Part 1 *Cells and Tissues.*   1971. Part 2, *Organs.* Edward Arnold, London.

Esau, K., 1965. *Plant Anatomy* (2nd edn), John Wiley & Sons Inc., New York.

Fahn, A., 1974. *Plant Anatomy* (2nd edn), Pergamon Press, London.

Foster, A. S. & Gifford, E. M., 1974. *Comparative Morphology of Vascular Plants* (2nd edn), W. H. Freeman & Co., San Francisco.

Books of informative scanning electron micrographs
Meylan, B. A. & Butterfield, B. G., 1972. *Three-Dimensional Structure of Wood*, Chapman & Hall, London.

Troughton, J. H. & Donaldson, L. A., 1972. *Probing Plant Structure*, Chapman & Hall, London.

Troughton, J. H. & Sampson, F. B., 1973. *Plants, A Scanning Electron Microscope Survey*, John Wiley & Sons, Australasia Pty., Ltd., Sydney.

3 Illustrated glossary

Abaxial: (surface) directed away from axis; e.g. lower surface of normal dorsiventral leaves.

Abscission zone: zone containing tissues which bring about abscission or cutting off of organs such as leaves, fruits or flowers

Accessory cell: subsidiary cell

Accessory transfusion tissue: transfusion tissue in mesophyll of some gymnosperm leaves rather than that related to vascular bundles

Acicular crystal: needle-shaped crystal

Acropetal: proceeding towards apex (e.g. of development)

Actinostele: protostele with xylem star-shaped in T.S.

Adaxial: (surface) directed towards axis; e.g. the upper surface of normal dorsiventral leaves

Adaxial meristem: meristematic tissue present adaxially in leaf, contributing to growth in thickness of petiole and midrib

Adnation: concrescence of organs or tissues of a different nature, e.g. a stamen and petal

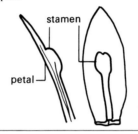

Adventitious organ: organ developing in an unusual position, e.g. roots at nodes of a stem or buds on root cuttings

Aerenchyma: parenchymatous tissue characterized by presence of large intercellular air spaces, cells sometimes stellate

Aggregate ray: in secondary vascular tissues, groups of small rays separated by fibres or axial parenchyma, but giving the appearance of a large ray

Albuminous cells: certain cells in phloem rays or phloem parenchyma of gymnosperms, related physiologically and situated adjacent to sieve elements; unlike companion cells, usually arising from

different cells from the sieve elements. Also applied to cells in certain seeds containing albumen

Aleurone grain: granules of reserve protein, present in many seeds

Aliform paratracheal parenchyma: *see* parenchyma

Amyloplastid, amyloplast: a leucoplast specialized to store starch

Anisocytic stoma: a stoma in which the three surrounding subsidiary cells(s) are of unequal sizes

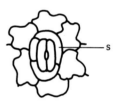

Annular thickening: secondary wall thickening deposited in rings on primary wall of tracheary elements (usually protoxylem)

Anomalous secondary growth: unusual type of secondary growth in thickness of an organ

Anomocytic stoma: a stoma in which the epidermal cells surrounding the guard cell pair (g) are not morphologically distinct from other epidermal cells

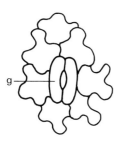

Anticlinal: usually referring to cell walls perpendicular to the surface of an organ

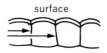

Antipodal cells: cells of the female gametophyte present at the chalazal end of the embryo sac in angiosperms

Aperture (of pollen grain): an area of characteristic shape in which exine is completely lacking or in which nexine alone is present; a pollen tube emerges via such an area. (Cf. stoma): the pore between a pair of guard cells.

Apex: distal portion of organ, e.g. root or shoot (or leaf)

Apical cell or apical initial: a cell which remains in the meristem, perpetuating. itself whilst dividing to form new cells which make up the body of the plant (in lower plants)

Apical meristem: a single cell or several layers of apical cells which are self-perpetuating and which by division in certain planes produce the precursors of the various tissues of the plant

Asterosclereid, astrosclereid: a branched sclereid

Atactostele: stele consisting of vascular bundles scattered throughout ground tissue, as in some monocotyledons. The apparently random arrangement in fact constitutes an orderly, if complex, pattern

Axial parenchyma: *see* parenchyma, xylem

Axial system: (i) all cells derived from fusiform cambial initials in secondary vascular tissues; (ii) cells elongated parallel to the long axis of an organ

Bark: non-technical collective term for tissues outside vascular cambium in secondarily thickened stems or roots

Basipetal: proceeding towards the base (usually of development)

Bifacial (dorsiventral) leaf: leaf with palisade parenchyma present on one surface of blade and spongy mesophyll on the other; having distinct dorsal and ventral surfaces

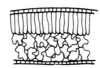

Body, primary: the parts of a plant developing from primary apical and intercalary meristems

Body, secondary: the parts of a plant made up of secondary vascular tissues and periderm, added to the primary body by the action of the lateral meristems cambium and phellogen

Brachysclereid, stone cell: short more or less isodiametric sclereid

Bulliform cell: enlarged epidermal cell common in leaves of Gramineae (as longitudinal rows of cells); sometimes called 'expansion cells', thought to bring about the unrolling of a developing leaf or 'motor cells' if involved with rolling and unrolling of leaves in response to water status of the leaves. Possibly merely water storage cells

Bundle cap: sclerenchyma or thick-walled parenchyma layer or layers of cells at phloem and/or xylem poles of vascular bundles

Bundle sheath: layer or layers of cells surrounding vascular bundles of leaves (and some stems); can be of parenchyma or sclerenchyma. May be of physiological significance in reducing water loss or acting as a boundary layer between bundle and other tissues; may contain chloroplasts of type associated with Kranz (C_3) metabolism

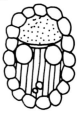

Bundle sheath extension: strip of ground tissue present along veins, extending between bundle sheath and epidermis or hypodermis abaxially or adaxially or both; consisting of parenchyma or sclerenchyma. Sometimes called a 'girder'. Often of characteristic outline in T.S. for a given genus or species

Callose: polysaccharide present in sieve areas, walls of pollen tubes, walls of fungal cells, etc.; often formed on injury of parenchymatous cells or as a rejection response on stigmatic hairs receiving pollen from a different species. On hydrolysis produces glucose

Callus: (i) layer of callose formed on sieve areas; (ii) tissue of parenchymatous cells formed as a result of wounding, or a tissue developing in tissue culture

Calyptrogen: in apical meristem of some roots, meristematic cells giving rise to the root cap; distinct from other apical meristematic cells forming the root itself

Cambial initials: self-perpetuating cells in vascular cambium and phellogen forming derivatives by periclinal division or increasing in number by anticlinal division. Fusiform initials give rise to axial phloem and xylem of secondary growth, and to ray initials. Ray initials give rise to rays in secondary phloem and xylem

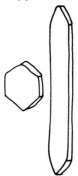

Cambium: (i) non-storied, composed of fusiform initials which, as seen in T.L.S. partially overlap one another in a random way and do not form horizontal rows; (ii) storied, composed of fusiform initials which, as seen in T.L.S. are arranged in horizontal rows

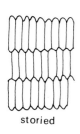 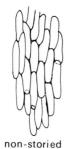

storied

non-storied

Cambium, vascular: a lateral meristem from which the secondary vascular tissues develop; (i) fascicular (f), the cambium forming within the vascular bundle; (ii)

interfascicular (i), the cambium forming between vascular bundles

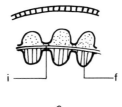

i ——————————— f

Cambial-like transition zone: a cytohistological zone visible in some shoot apices

Caruncle: a fleshy outgrowth of the integuments at the micropylar region of a seed

Casparian strip, band: a band-like structure (c) in the primary wall containing lignin and suberin. Particularly characteristic of endodermal cells of roots where the band is present in radial and transverse anticlinal walls; similar cells are sometimes observed in stems, between cortex and stele, also in exodermis cells of some roots. In ferns, individual vascular bundles can be enclosed by such an endodermis. Their function is of physiological significance, and they serve to reduce or prevent water movement through the cell walls in a radial direction

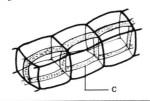

c

Cell: structural unit of living organism; living plant cells consist of wall and protoplast

Cell plate: the part of a cell wall developing between the two daughter nuclei in telophase of cell division

Cellulose: carbohydrate consisting of long chain molecules comprising anhydrous glucose residues as basic units; a principle constituent of plant cell walls

Central mother cells: cytohistological zone of shoot apex in the region below the surface layers; commonly used in describing gymnosperm apices

Chalaza: region in ovule where integuments and nucellus connect with the funiculus

Chimera: a combination in a single plant organ of cells or tissues of different genetic composition

Chlorenchyma: a specialized parenchymatous tissue containing chloroplasts, e.g. palisade mesophyll, spongy mesophyll

Chlorophylls: the green pigments present in chloroplasts

Chloroplast: a specific protoplasmic body in which photosynthesis takes place; usually disc-shaped

Chromoplast: plastid, containing pigment (*see* plastid)

Cicatrice: scar left by separation of one part or organ from another; e.g. hair base of deciduous hair

Coencyte: group of protoplasmic units, a multinucleate structure. In angiosperms usually refers to multinucleate cells

Coleoptile: sheath surrounding apical meristem and leaf primordia of grass embryo

Coleorrhiza: sheath surrounding radicle of grass embryo

Collenchyma: (i) general – supporting or mechanical tissue in young organs and primary structures such as certain leaves. Consisting of cells with uneven, mainly cellulosic wall thickenings; (ii) angular – cell wall thickenings concentrated at angles; (iii) lacunar – cells with characteristic intercellular spaces and wall thickenings opposite the spaces; (iv) lamellar – wall thickenings mainly on anticlinal (tangential) walls

Colleter: multicellular glandular hair with stalk and head having a sticky secretion

Columella: (i) in some roots central portion of root cap in which cells are arranged in longitudinal files; (ii) in other usage, means a small pillar

Companion cell: specialized parenchyma cell (c) associated with and derived from the same mother cell as the sieve tube element (member), to which it is physiologically connected

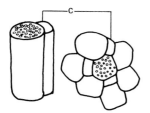

Complementary cells: loose tissue formed towards periphery by phellogen of the lenticel; cell walls may or may not be suberized

Compression wood: reaction wood in conifers formed on lower side of branches etc.; dense in structure with strong lignification of tracheid walls

Cork: phellem

Cork cell: (i) dead cell arising from the phellogen, whose walls are impregnated with suberin; function frequently protective; (ii) in an epidermis, a short cell with suberized walls; characteristic of grasses

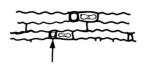

Corpus: the cells below the surface layer(s) (tunica) of angiosperm shoot apex in which cell divisions take place in various planes, giving rise to increase in apex volume (tunica-corpus theory)

Cortex: region of ground tissue between epidermis or its secondary replacement and vascular cylinder of axis

Crassulae: transversely-oriented thickenings in tracheid walls of gymnosperms accompanying the pit pairs and formed by the intercellular material of primary wall layers. Also called bars of sanio

Cross field: area formed by the walls of a ray cell and an axial tracheid as seen in R.L.S.; mainly used in description of conifer woods; cr = cross field pit, t = tracheid

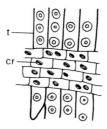

Crystal: cell inclusion, usually of calcium oxalate, exhibiting a range of forms; sometimes of taxonomic or diagnostic significance

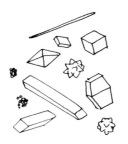

Cuticle: layer of cutin, a fatty substance which is almost impermeable to water; present on outer walls of epidermal cells, sometimes extending into supra- and sub-stomatal cavities as a very thin lining

Cuticle layer: outer portions of the epidermal cell walls impregnated with cutin

Cutinization: process of deposition of cutin in cell walls

Cylinder, central or vascular: that part of the axis of a plant consisting of vascular tissue and the associated ground tissue. Equivalent to term 'stele' but lacking evolutionary implications

Cystolith: a specific outgrowth of the cell wall on which calcium carbonate is

deposited; characteristic of certain families, e.g. Moraceae

Cytochimera: in a single plant organ, a combination of cells which are of different chromosome number

Dedifferentiation: reversal of differentiation; usually occurs in more or less mature parenchymatous cells which secondarily assume meristematic activity

Dermal tissue: epidermis or periderm

Dermatogen: meristem forming epidermis

Diacytic stoma: a stoma in which the guard cell pair has one subsidiary (s) cell at either pole

Diaphragm: a partition of cells in an elongated air cavity in an organ; may be transverse or longitudinal

Diarch: primary root with two protoxylem strands (and poles)

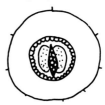

Dictyostele: a siphonostele in which the leaf gaps are large and partly overlap one another so as to divide the stele into separate bundles in each of which the phloem surrounds the xylem

Diffuse porous wood: secondary xylem in which there is no clear size distinction between vessels formed at beginning and end of a season's growth

Dorsiventral leaf: bifacial leaf

Druse: a compound crystal more or less spherical in shape and in which the many component crystals protrude from the surface

Duct: an elongated space formed schizogenously (q.v.), lysigenously (q.v.) or schizo-lysigenously and which may contain secretions or air

Ectocarp: the outermost layer of the pericarp (fruit wall)

Elaioplast: an oil-producing and storing leucoplast

Elaiosome: an outgrowth on a fruit or seed that contains large oil-storing cells

Emergence: a projection of the surface of a plant organ consisting of epidermal cells and cells derived from underlying tissues

Endarch xylem: a primary xylem strand in which the first-formed elements are closest to the centre of the axis, as in the shoots of most spermatophyta

Endocarp: the innermost layer of the pericarp (fruit wall)

Endodermis: layer of ground tissue forming sheath or cylinder around the vascular region and whose cell walls bear casparian strips; usually readily recognizable in roots, sometimes less easily defined in stems, at inner boundary of cortex

Endodermoid layer: layer of cells surrounding central vascular cylinder of stem, in position of endodermis, but in which casparian strips are not distinguishable. (The distinction between endodermis and endodermoid layer is not recognized by some)

Endogenous: developing from internal tissues, as for example a lateral root

Endosperm: a nutrient tissue formed within the embryosac of the spermatophyta

Endothecium: a layer of cells situated below the epidermis in the pollen-sac wall having characteristic wall thickenings

Epiblast: a small growth present opposite the scutellum in the embryo of some Gramineae

Epiblem: a term used for the outermost layer (epidermis) of primary roots

Epicarp: *see* Exocarp

Epicotyl: the stem of an embryo and seedling above the cotyledons

Epidermis: the outermost cell layer of primary tissues of the plant; sometimes comprising more than one layer – multiseriate epidermis or multiple epidermis

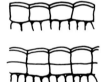

Epithelial cell: cell lining a cavity or canal; usually with secretory function

Epithem: the tissue between the vein ending and the secretory pore of a hydathode

Ergastic matter: the non-protoplasmic products of the metabolic processes of the protoplasm; includes starch grains, oil droplets, crystals, tannins and certain liquids; found in the cytoplasm, vacuoles and cell walls

Eustele: thought of as the most advanced type of stele in the phylogenetic sense in which the vascular tissue forms a hollow reticulate cylinder made up of collateral or bicollateral vascular bundles

Exalbuminous seed: a seed lacking endosperm when mature

Exarch xylem: a strand of primary xylem in which the first-formed elements are furthest from the centre of the axis, as in roots of spermatophyta

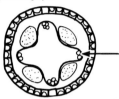

Exine: the outer wall of a mature pollen grain

Exocarp, epicarp: the outermost layer of the pericarp (fruit wall)

Exodermis: present in some roots as modified layer or layers of cells of the outermost part of the cortex, the walls of which are more or less thickened and contain lamellae of suberin

Exogenous: developing from outer tissues, as for example, an axillary bud

Expansion cell: bulliform cell

Fascicular: part of, or situated within, a vascular bundle; *see* cambium fascicular

Fibre: an elongated sclerenchymatous cell often with tapered ends; walls with secondary thickening, usually of lignin. With or without a living protoplast when mature. Loose usage to mean wood elements in general

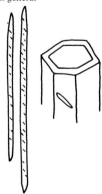

Fibre, gelatinous: xylem fibre in which inner layers of secondary wall can swell on the absorption of water

Fibre, libriform: fibre in secondary xylem, usually with few, simple pits

Fibre, pericyclic: fibre found in the outer region of the vascular system, either in the phloem, as a phloem fibre or outside it and termed a perivascular fibre

Fibre, septate: a fibre with thin transverse septa formed after laying down of secondary walls

Fibre-sclereid: a cell with characters intermediate between a fibre and a sclereid or a lignified fibre developing from parenchyma in non-functional phloem

Fibre-tracheid: a cell with characters intermediate between a libriform fibre and a tracheid, commonly with pointed ends; pits bordered, with slit-like apertures; found in secondary xylem

Funiculus: stalk attaching an ovule to the placenta

Fusiform: elongated, with pointed ends, as in fusiform cambial initial

Gap, branch: a parenchymatous region in a siphonostele above the position where a

branch trace connects with the vascular cylinder of the stem (b)

Gap, leaf: a parenchymatous region in a siphonostele above the position where a leaf trace connects with the vascular cylinder of the stem (l)

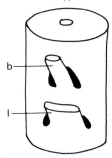

Ground tissue: tissue in a stem or root other than epidermis, periderm or vascular system

Growth ring: increment of secondary xylem or phloem formed during a period of growth; since there may be more than one increment in a year, the term 'annual ring' should be used with caution

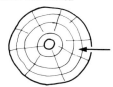

Guard cells: a pair of specialized epidermal cells bordering a pore and constituting a stoma; changes in shape of the guard cells effect the opening or closing of the pore

Gum: a non-technical term applied to some of the materials arising from break-down of certain components of plant cells

Haplostele: prostele with a more or less circular cross-section to the xylem

Hardwood: general term for secondary xylem of angiosperms

Heartwood: inner part of wood of a trunk or branch which has lost the ability

to conduct water; generally darker than sapwood because of the materials deposited in it

Helical wall thickening, 'spiral' wall thickening: secondary or tertiary wall material deposited on a primary or secondary wall respectively in certain tracheary elements

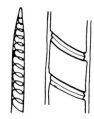

Heterocellular ray, heterogeneous ray: ray in secondary vascular tissues composed of more than one form of cell. In dicotyledons these are all parenchymatous cells; in gymnosperms, tracheids or radial resin canals may be present with the parenchymatous cells; radial canals occur in some angiosperms

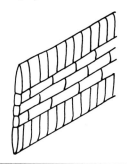

Hilum : (i) the funiculus scar in seeds; (ii) the portion of a starch grain acting as a nucleus around which the layers are deposited

Histogen: in an apical meristem, a layer or layers of cells which develop into one of the three systems of the organ: dermatogen (d) → epidermis; periblem (pe) → cortex; plerome (pl) → vascular system. The number of layers of cells in each can vary from species to species and within a single species, and there may be 2 or 4 histogens in some plants

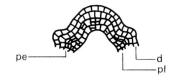

Homocellular ray, homogeneous ray: ray in secondary vascular tissue composed of one (parenchymatous) cell form only

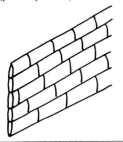

Hydathode: structure through which liquid water can be extruded, sometimes glandular; found mainly on leaves; thought to be modified stomata

Hypodermis: layer or layers of cells immediately below the epidermis, not derived from the same initials as the epidermis (as can be seen by lack of coincidence of anticlinal walls of epidermis and hypodermis), differing in appearance from tissues below them. The root exodermis is a specialized hypodermis

Idioblast: a cell clearly distinguishable from others in the tissue in which it is embedded, in size, structure or content; e.g. sclereid, or tanniniferous idioblast, or crystal containing cell

Initial: (i) *see* apical initial; (ii) a meristematic cell which differentiates into a specialized element; (iii) marginal, cells along growing lamina of leaf which contribute cells to the protoderm

Integument: enveloping layer surrounding the nucellus

Intercellular space: a space between cells of a tissue; may arise by (a) splitting apart of cells along the middle lamella (schizogenous) or by (b) dissolving cells (lysigenous) or (c) by tearing apart of cells (rhexigenous)

Interfascicular: (ground tissue) between vascular bundles in primary stem (primary medullary rays)

Internode: part of a stem between two nodes

Intervascular pitting: pitting between tracheary elements

Interxylary: within or surrounded by xylem, e.g. interxylary cork, cork developing amongst elements of xylem tissue, or interxylary phloem

Intine: the inner wall of a mature pollen grain

Intraxylary: on the inner side of the xylem

Intrusive growth: type of cell growth in which cells grow in length or width between other cells, separating them at the middle lamella; a common form of growth extension of some fibres

Isobilateral leaf, isolateral leaf: leaf with palisade tissue on both surfaces of blade

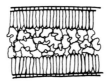

Lacuna: space between tissues, usually filled with air; see intercellular space

Lamina: blade, or expanded part of leaf

Laticifer: cell or series of cells with characteristic latex fluid content; usually tubular in shape; may be branched or unbranched

Laticifer, articulated: compound laticifer, formed of longitudinal series of cells, with walls between cells entire or perforated

Laticifer, non-articulated: single cells which may be coenocytic and branched but are not joined to form long tubes

Laticiferous cell: non-articulated or simple laticifer

Laticiferous vessel: articulated laticifer with walls between adjacent cells perforated

Leaf buttress: initial stage of development of leaf primordium in apical meristem

Lenticel: part of periderm in which cells are loosely packed and may or may not be suberized

Leucoplast: a colourless plastid

Lignification: impregnation with lignin

Lignin: an organic complex of high carbon-content substances, distinct from carbohydrates; present in matrix of cell walls of many cells

Lithocyst: a cell containing a cystolith

Lumen: the internal space bounded by a cell wall in a plant cell

Lysigenous: form of separation by dissolving (usually enzymatically) of cells

Mechanical tissue: cells with more or less thickened walls, e.g. collenchyma of the primary growth and sclerenchyma of primary and secondary growth. Also called supporting tissue

Medulla: pith

Medullary bundle: vascular bundle located in the pith or central ground tissue

Medullary ray: *see* ray

Meristele: one of the bundles of a dictyostele; see vascular bundle

Meristem: tissue which by division produces new cells which undergo differentiation to form mature tissue and at the same time frequently perpetuates itself

Meristematic cell: a constituent cell of a meristem; shape, degree of wall thickness and extent of vacuolation varies in cells found in different meristematic regions

Meristem, apical: a meristem at the apex of shoot or root which by division gives rise to cells forming the primary tissues of shoot or root.

Meristem, axillary: meristem located in leaf axil; capable of giving rise to axillary bud

Meristem, ground: meristematic tissue originating in an apical meristem, producing tissues other than epidermis and vascular tissues

Meristem, intercalary: meristematic tissue derived from apical meristem which during the course of development of the plant becomes separated from it by regions or more or less mature tissues

Meristem, lateral: meristem parallel to the circumference of the plant organ in which it occurs, e.g. cambium phellogen

Meristem, marginal: meristem located along the margin of a leaf primordium and forming the leaf blade

Meristem, mass: meristematic cells which divide in various planes and contribute to increase in tissue volume

Meristem, plate: parallel-layered meristem with planes of cell division in each layer perpendicular to the surface of the organ, which is usually a flat one

Meristem, primary thickening: lateral meristem derived from the apical meristem and responsible for the primary increase in width of the shoot axis; commonly found in monocotyledons

Meristem, rib: (i) one of the regions of the shoot apex; (ii) a meristem composed of parallel series of cells in which transverse divisions characteristically take place

Mesarch xylem: the condition in which protoxylem in a primary xylem strand develops first in the centre of the strand and continues to develop both centrifugally and centripetally, e.g. in shoots of ferns

Mesocarp: the central region of a fruit wall (pericarp)

Mesocotyl: internodal region between scutellar node and coleoptile in the embryo and seedling of a grass

Mesomorphic: refers to the structural features normally found in plants adapted for growth in conditions of adequate soil water and a fairly humid atmosphere (mesophytes)

Mesophyll: chlorenchymatous and other parenchymatous tissues of the leaf blade contained brtween the epidermal layers

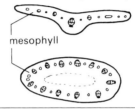

mesophyll

Mesophyll, plicate: compact mesophyll cells in which the cell walls have inwardly directed projection or folds

Mesophyte: *see* mesomorphic

Mestome sheath (bundle sheath, if cells thick walled): a sheath of cells with thickened walls, surrounding a vascular bundle; frequently endodermoid

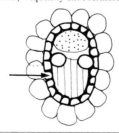

Metaphloem: primary phloem which develops after the formation of protophloem and before secondary phloem (should that also develop)

Metaxylem: primary xylem which develops after the formation of the protoxyclem and before secondary xylem (should that also develop)

Microfibril: submicroscopic, thread-like usually cellulosic components of plant cell walls

Micropyle: a small opening between the integuments at the free end of an ovule

Microspore: male spore from which develops the male gamete

Microsporocyte: a cell which develops into a microspore

Middle lamella: material, usually pectic, acting rather like a cement, present between the primary walls of adjacent cells

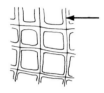

Mitochondrion: minute body in cytoplasm, containing respiratory enzymes; also called a chondriosome

Mother cell: a cell which on division gives rise to other cells and thus loses its identity; e.g. a guard cell mother cell

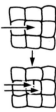

Motor cell: *see* bulliform cell

Mucilage cell: cell containing a mucilage, gum or similar carbohydrate material

Mucilage duct: duct containing mucilage, gum or similar carbohydrate material

Multiseriate: consisting of many layers of cells (*see also* ray, multiseriate)

Nectary: multicellular, glandular structure capable of secreting sugary solution. Floral nectaries occur in flowers; extrafloral nectaries occur in other plant organs

Nexine: the inner layer of the exine of a pollen grain

Node: a region of a stem where a leaf or leaves or buds are attached; a rather loose term since the division between node and internode is rarely distinct

Node, unilacunar: node with one leaf gap in relation to each leaf

Node, multilacunar: node with more than one leaf gap in relation to each leaf (usually used when 4 or more gaps are present)

Nucellus: tissue within ovule in which the female gametophyte develops

Osteosclereid: a 'bone-shaped' sclereid

Paracytic stoma: stoma in which the guard cells have one or more subsidiary cells adjacent to and parallel to them on either flank

Parenchyma: living cells, frequently thin-walled but sometimes, particularly in the xylem, with lignified, thickened walls; varying in shape and size according to position in the plant

Parenchyma, apotracheal: axial parenchyma of the secondary xylem, typically not associated with vessels. (i) banded, concentric uni- or multiseriate bands, sometimes complete rings as seen in T.S.; (ii) diffuse, single cells as seen in T.S., distributed irregularly among fibres (usually in axial chains of 4 cells); (iii) diffuse in aggregates; (iv) initial, bands produced at beginning of growth ring; (v) terminal, bands produced at end of growth ring

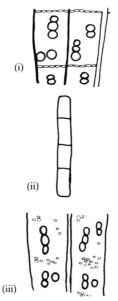

Parenchyma, paratracheal: axial parenchyma of secondary xylem associated with vessels or tracheids. (i) aliform; (ii) confluent; (iii) scanty – incomplete sheath round vessel; (iv) vasicentric – complete sheath of variable width round individual or groups of vessels

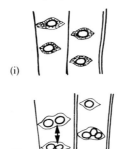

Parenchyma, xylem: parenchyma occurring in secondary xylem (i) axial; (ii) radial (of rays)

Parthenocarpy: development of ovule to fruit without fertilization (e.g. cultivated bananas)

Passage cell: thin-walled cell in root or stem endodermis or exodermis, conspicuous because of thickened walls of other endodermal cells; casparian strips present in walls if in an endodermis

Pectic compounds: polymers of galacturonic acid and its derivatives; main constituent of middle lamella and intercellular substances, also a component of cell walls

Perforation plate: perforated end cell wall of vessel element; generally in end walls. (i) simple, surrounded by rim only; (ii) scalariform, several to numerous elongated pores with bars between them (ladder-like); (iii) reticulate, net-like; (iv) foraminate, numerous more or less circular pores

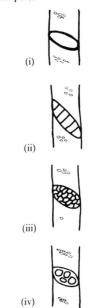

Periblem: the meristem forming the cortex, according to Hanstein's system

Pericarp: fruit wall developed from the ovary wall

Periclinal: parallel to the surface of an organ

Pericycle: ground tissue; in angiosperms usually present in roots between endodermis and vascular tissue, but less common in stems and probably best not applied to them.

Periderm: secondary protective tissue replacing epidermis in stems or roots which exhibit secondary growth in thickness: consisting of phellogen producing by periclinal division phellem (cork) to the outside and phelloderm to the inside

Perisperm: nutrient tissue present in some seeds; originating from the nucellus

Phellem: cork

Phelloderm: layer or layers of cells produced to the inner side of the phellogen by periclinal division

Phellogen: cork cambium, a secondary lateral meristem, producing phelloderm internally and phellem externally; may be superficial (arising at or close to epidermis) or deep-seated (arising in deeper cortical or phloem layers)

Phloem: main tissue which translocates assimilated products in vascular plants; composed mainly of sieve elements and companion cells (or albuminous cells), parenchyma, fibres and sclereids. Phloem, included (interxylary): secondary phloem embedded in the secondary xylem of some dicotyledons. Phloem, internal (intraxylary): primary phloem present on the inner side of the primary xylem

Phyllotaxy: the mode of arrangement of leaves on the axis of a stem

Pillar cell: (i) description of subepidermal sclereids in the seed coat of some Leguminosae; (ii) in Restionaceae, specialized, lignified cells of the stem parenchyma sheath extended to the

epidermis and dividing the chlorenchyma into longitudinal channels

Piliferous cell: cell, usually of the epidermis, bearing a hair or trichome

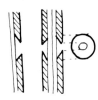

Pit: a thin area of a secondarily thickened cell wall consisting of middle lamella and primary wall only

Pits, alternate: in tracheary elements, pits arranged in diagonal rows as seen in T.L.S. or R.L.S.

Pit, bordered: a pit in which the aperture is smaller than the pit membrane and in which the secondary wall overarches the pit membrane and pit cavity

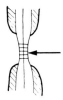

Pit, half-bordered: a pit pair in which the aperture is bordered on one side of the middle lamella and not bordered on the other side

Pits, opposite: in tracheary elements, pits arranged in horizontal pairs or short horizontal rows as seen in T.L.S. or R.L.S.

Pit, simple: a pit in which the aperture and pit membrane are similar in size

Pit, vestured: bordered pit with projections, either simple or branched, on secondary wall forming border of pit chamber or aperture

Pith: central ground tissue of stem and root; often parenchymatous, sometimes sclerotic or containing sclereids or other cell types

Placenta: region of attachment of the ovules to the carpel

Placentation: position of the placenta in the ovary

Plasmalemma: membrane forming the outer surface of the cytoplasm

Plasmodesma: thin cytoplasmic strand passing through a pore in the cell wall usually connecting the protoplasts of two adjacent cells

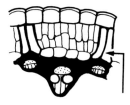

Plastid: protoplasmic body separated from the cytoplasm by a membrane

Plastochron: period of time between the initiation of two successive, repetitive phenomena, e.g. between the initiation of two leaf primordia

Plectostele: protostele in which xylem is arranged in longitudinal plates which may be interconnected

Plerome: the meristematic cells of an apex responsible for the formation of the primary vascular system, its parenchyma and pith (if present), according to Hanstein's theory

Plumle: bud or shoot apex of the embryo

Pneumatode: a group of cells present in a velamen, with spiral secondary wall thickenings; may also be used for other aerating tissue

Pneumatophore: negatively geotropic, aerial root projection formed on certain species growing in swampy ground, e.g. *Taxodium*; serves for gas exchange

Pollen tube: projection of vegetative cell of pollen grain, occurring on germination of the grain, covered by intine only

Pollinium: a mass of pollen grains adhering together and usually dispersed as a unit

Polyarch: primary xylem of root with a large number of protoxylem strands; e = endodermis, pl = phloem, px = protoxylem, mx = metaxylem

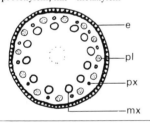

Polyderm: protective tissue consisting of alternating bands of endodermis-like cells and non-suberized parenchyma cells

Pore: in wood, an unscientific term for the vessel elements as seen in cross-section

Primary cell wall: the outermost layer of the wall of the cell, in which cellulose microfibrils show apparently random orientation; such structures are beyond the resolving power of the light microscope so primary wall is usually taken to mean the wall laid down during the phase of growth in size of the cell. In fibres with apical intrusive growth, areas of secondary wall thickening are laid down before the cells have completed their growth in size

Primary pit field: a thin area of the primary wall with a concentration of pores and plasmodesmata

Primary phloem: phloem tissue developed from procambial tissue in the primary growth zones of the plant; protophloem forms first (pp), then metaphloem (mp); rays are absent

Primary xylem: xylem tissues developed from procambial tissue in the primary growth zones of the plant; protoxylem forms first (px), then metaxylem (mx); (rays are absent from the primary xylem itself, but may be present between vascular bundles)

Primordium: the earliest stage of differentiation in an organ, group of cells or single cell; e.g. root primordium

Procambium: a primary meristem differentiating to form the primary vascular tissues

Proembryo: an embryo at its earliest stages of development before the start of organ differentiation

Promeristem: in an apical meristem, the initial cells and their immediate derivatives

Prophyll: one of the earliest leaves of a lateral branch

Proplastid: a plastid in the earliest stages of development

Prosenchyma: elongated parenchyma cells with thickened lignified walls; often fibre-like

Protoderm: the meristem developing into an epidermis

Protophloem: the elements of primary phloem which develop first

Protoplast: a living cell unit

Protostele: a stele in its simplest form consisting of a solid central xylem cylinder surrounded by phloem

Protoxylem: the elements of primary xylem which develop first

Provascular bundle: procambial strand; *see* procambium

Pseudocarp: a false fruit in which floral organs other than carpels participate in forming the fruit wall, e.g. apple

Pulvinus: the swelling at the base of a leaf petiole or leaflet petiolule

Radicle: the embryo root

Ramiform pit: simple pit in which canals are coalescent

Raphe: a ridge on a seed formed by that part of the funiculus which was fused to the ovule

Raphide: a needle-shaped crystal of calcium oxalate; usually one of a number of crystals arranged parallel to one another in a mucilaginous sac or raphide sac

Ray, heterocellular: *see* heterocellular ray

Ray, homocellular: *see* homocellular ray

Ray, pith: parenchymatous interfascicular region of stem

pith ray

Ray, vascular: a tissue system oriented radially through the secondary xylem (xylem ray) and secondary phloem (phloem ray) and derived from the cambial ray initials; ca=cambium, co=cortex, p=phloem, r=ray, x=xylem

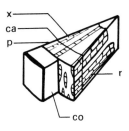

Ray, uniseriate: ray of secondary vascular tissue one cell wide

Ray, multiseriate: ray of secondary vascular tissue 2 to many cells wide as seen in T.L.S.

Ray tracheid: tracheid occurring in the radial system of the wood of some conifers, usually at the ray margins

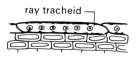

ray tracheid

Reaction wood: wood with distinctive features forming in leaning or crooked branches and twigs; termed tension wood in angiosperms and compression wood in conifers

Resin duct or canal: schizogenous duct containing resin

Reticulate cell, wall thickening: secondary wall thickening in tracheary elements with a net-like appearance

Reticulate venation: leaf blade veins forming an anastamosing, net-like system

Rhexigenous: formed by tearing apart of cells (*see* intercellular space)

Rhytidome: the outer part of the bark composed of the periderm and all tissues external to it

Ribosome: minute protoplasmic organelle containing messenger RNA; concerned with protein synthesis

Ring porous wood: secondary xylem in which vessels produced at the start of a

season's growth are distinctively larger than those formed later in the season

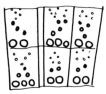

Root, contractile: specialized root capable of contraction; helps to maintain a plant or part of a plant at the correct depth in the soil

Root cap: cells cut off by the calyptrogen in the root apical meristem and forming a protective cap cushioning the apex itself

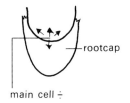
rootcap

main cell ÷

Root hair: a type of trichome developed from the root epidermis; maybe short-lived, it absorbs solutions from the soil

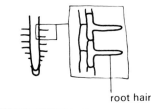
root hair

Sapwood: outer part of xylem of a tree or shrub containing living cells and reserve materials

Scalariform: ladder-like; a closely parallel arrangement of structures in the cell wall of an element, e.g. secondary wall thickening or perforation plate

Scale: a flattened type of trichome attached along or near to one edge

Schizogenous: formed by separation or splitting, usually refers to intercellular

spaces which originate by cells parting at the middle lamellae

Schizo-lysigenous: formed by separation and dissolution of cells, usually refers to intercellular spaces

Sclereid: a form of sclerenchymatous cell with lignified walls, usually relatively short; a range of types exists:
  astrosclereid, branched or ramified sclereid
  brachysclereid or stone cell, short, more or less isodiametric sclereid
  fibre sclereid, a cell intermediate in length between a fibre and sclereid
  macrosclereid, elongated sclereid with uneven secondary wall thickening; when present in testa of leguminous seeds also called a malpighian cell
  ostoesclereid, 'bone-shaped' sclereid
  trichosclereid, hair like sclereid

Sclerenchyma: a mechanical or supporting tissue of cells with lignified walls, made up of fibres, sclereids and fibre-sclereids

Sclerification: the process of changing into sclerenchyma by the progressive lignification of secondary walls

Scutellum: part of the embryo in Gramineae

Secondary cell wall: cell wall deposited after the primary cell has ceased growth in volume (*see* primary cell wall); microfibrils as seen with the TEM have an ordered more or less parallel orientation in each definable layer

Secretory cavity: a cavity filled with the breakdown products of cells which formed the cavity

Secretory cell: a specialized living cell which secretes or excretes substances

Secretory duct: duct formed schizogenously, frequently lined by thin-walled secretory epithelial cells which secrete substances into the duct

Separation layer: the layer or layers of cells which disintegrate in the abscission zone

Sexine: the outer layer of the exine of the pollen grain

Shoot: the stem and its appendages

Sieve area: an area of the wall of a sieve element which contains a concentration of pores, each callose-lined and encircling a strand of protoplasm which connects the protoplast of one sieve element with that of the next

Sieve cell: a sieve element with relatively undifferentiated sieve areas; narrow pores and connecting strands; found in gymnosperms and lower vascular plants

Sieve element: phloem element whose main function is the axial transport of assimilates, comprising (i) sieve cells and (ii) sieve tube elements or members

Sieve plate: specialized areas in the walls of sieve tube members

Sieve tube: a series of sieve tube elements or members joined together end to end and connected by sieve plates

Silica body: opaline cell inclusion; the shape of a silica body may be characteristic for a family or group within a family

Silica cell: (i) a cell containing 1 or more silica bodies, (ii) an epidermal cell containing a silica body

Siphonostele: a stele composed of a hollow cylinder of vascular tissue with a central pith; (i) amphiphloic – phloem both to interior and exterior of xylem; (ii) ectophloic – phloem to exterior of xylem cylinder

Soft wood: common name for gymnosperm wood, particularly that of the Coniferae. Some gymnosperm wood can, in fact, be very hard

Solenostele: amphiphloic siphonostele in which successive leaf gaps are well separated from one another

Spiral thickening: *see* helical thickening

Starch: an insoluble carbohydrate acting as one of the commonest storage products of plants, composed of anhydrous glucose residues

Starch grain: a cell inclusion composed of starch; frequently with a characteristic shape for a particular species or group of species. The radiating chain structure of the crystalline residue produces a characteristic 'Maltese Cross' when the grain is viewed between crossed polars in the microscope

Starch sheath: name given to the innermost layer of the cortex if it is specialized to store starch; probably homologous with the endodermis

Stele: the part of the plant axis made up of the primary vascular system and its associated ground tissue

Stele, polycyclic: a stele with two or more concentric circles of vascular tissue

Stellate: star-shaped; *see* aerenchyma and sclereids

Stereome: all mechanical tissue of the plant

Stoma: a pore in the epidermis encircled by two guard cells; often used to describe both the pore itself and the two guard cells which surround it and regulate its size

Storied (or storeyed) **tissue**: tissue in which the cells are arranged in horizontal series as seen in T.L.S. and R.L.S. e.g. storied cambium which gives rise to storied xylem and phloem; storied rays may be apparent even when other tissues have lost their regular arrangement during

growth adjustments. Produces ripple marks in wood figure

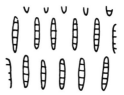

Stroma: the structural framework of a plastid

Stone cell: *see* brachysclereid under sclereid

Styloid: crystal with elongated prismatic shape; ends flat or pointed

Suberization: the process of deposition of suberin in cell walls; suberin is similar in nature to cutin

Subsidiary cells: epidermal cells which together with the guard cells make up the stomatal apparatus. Subsidiary cells are frequently distinguishable from other epidermal cells by their shape or wall thickness. Occasionally they can only be discerned by developmental studies

Surface layer, surface meristem: a histological zone in the gymnosperm apex

Suspensor: the connection between the main part of the embryo and the basal cell; may have a function in nutrition of an embryo

Syncarpy: fusion of carpels in a flower and ovary

Synergids: cells in the mature embryo sac present alongside the egg cell

Tangential: at right angles to a radius; tangential walls are frequently periclinal

Tangential longitudinal section (T.L.S.): a section made at a tangent to and parallel with the long axis of a cylindrical organ

Tannin: a collective term used for a range of polyphenolic substances deposited in certain plant cells; common, for example, in bark, whence it is extracted for tanning

Tapetum: innermost layer of cells of pollen sac wall; their contents nourish developing pollen grains and also provide part of the protein involved in recognition systems between pollen and stigma; (i) amoeboid – a tapetum in which the protoplasts of its cells penetrate between the pollen mother cells; (ii) glandular – a tapetum in which the cells disintegrate in their original position

Tension wood: reaction wood formed on the upper side of branches or leaning or bent stems of angiosperms; fibres characteristically gelatinous and little lignified

Tertiary wall: wall thickening to inner side of secondary wall

Tepal: a perianth member in flowers that lack distinction between petals and sepals, e.g. *Tulipa*

Testa: the seed coat

Tetrarch: primary xylem of a root with four protoxylem strands

Tissue, conjunctive: (i) a special type of parenchyma associated with included phloem in dicotyledons having anomalous secondary thickening; (ii) parenchyma present between secondary vascular bundles in monocotyledons having secondary thickening

Tissue, expansion: an intercalary tissue in the outer part of the inner bark, formed mainly by the phloem rays, which enables the bark to expand without tearing

Tissue, ground: all tissues of mature plants except the epidermis, periderm and vascular tissues

Tissue, mechanical: supporting cells, e.g. sclerenchyma and collenchyma

Tissue, transfusion: tissue surrounding or associated with vascular bundles in the leaves of gymnosperms, composed of tracheids and varying amounts of parenchyma cells

Tonoplast: membrane of the cytoplasm where it borders a vacuole

Torus: central thickening of pit membrane in a bordered pit of certain gymnosperms;

made up of middle lamella and primary wall material

Trabecula: bar-like projection of cell wall crossing a cell lumen

Trace: (i) branch – vascular system joining the main stem and vascular supply of a branch; (ii) leaf – vascular system joining the main stem to the leaf vascular system

Tracheary element: xylem element involved in water transport; includes vessel elements, tracheids and tracheoidal vessel elements. A useful term used to describe water conducting tissue when the exact cell type has not been ascertained

Tracheid: an imperforate tracheary element, i.e. with intact pit membrane between it and adjacent elements

Tracheoidal vessel element: a perforated vessel element which is very elongated and looks in all other respects like a tracheid

Transfer cell: parenchymatous cell with minute ingrowths of cell wall; concerned with movement of materials e.g. in seedlings

Transfusion tissue: *see* tissue, transfusion

Transverse section (T.S.): a cross-section

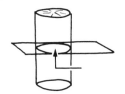

Traumatic tissue: wound tissue, e.g. callus or resin filled cavities of a traumatic resin duct

Triarch: a primary root with xylem exhibiting three protoxylem poles in T.S.

Trichoblast: specialized cell in root epidermis which gives rise to a root hair

Trichome: epidermal appendage, includes hairs, scales and papillae; may be glandular or non-glandular

Tunica: outermost layer or layers of cells in apical meristem of angiosperm shoot in which most cell divisions are anticlinal; in the corpus, cell divisions are anticlinal, periclinal and also in other planes

Tylosis: intrusion of a ray or axial parenchyma cell into a vessel element lumen by perforation of a pit membrane; may or may not be lignified; occur rarely in tracheids

Tylosoid: an epithelial cell which proliferates in an intercellular cavity such as a resin duct

Unifacial leaf: a leaf which develops from one side of the leaf primordium only and in consequence has only an encircling adaxial or abaxial epidermis (may be secondarily flattened and appear dorsiventral)

Uniseriate: cells in one layer, e.g. uniseriate ray

Upright ray cell: cell of rays of secondary vascular tissues; longer axially than radially

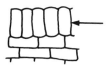

Vacuolation: vacuole formation

Vacuole: a volume enclosed in the cytoplasm separated from it by the tonoplast, containing cell sap

Vascular: referring to the xylem or phloem or both

Vascular bundle: an organized strand of conducting tissue composed of xylem and phloem, and in most dicotyledonous stems, cambium; (i) closed – lacking cambium, as in monocotyledons; (ii) open – with cambium (c); (iii) collateral – with one phloem and xylem pole; (iv) bicollateral – with two phloem poles, one at either end of the xylem pole, but with only one cambial zone; (v) concentric – (a) amphicribal, xylem surrounded by phloem and (b) amphivasal, phloem surrounded by xylem

Vascular cambium: lateral meristem forming secondary vascular tissues

Vein: a vascular bundle or group of closely parallel bundles in a leaf, bract, sepal, petal, or flat stem

Velamen: multiseriate epidermis, a characteristic tissue of many aerial roots in Orchidaceae and Araceae; may be present in some terrestrial roots

Venation: the arrangement of veins in an organ; (i) closed vein endings which anastome in a leaf blade; (ii) open vein endings which are free, i.e. do not anastome in a leaf blade

Vessel: a series of perforated vessel elements joined end to end

Vessel element: a tracheary member of a vessel

vessel elements

Wall, tertiary layer: a layer of thickening material to the inner side of a secondary wall; e.g. tertiary spiral wall thickening

Wart: fine granular protrusions on inner surface of secondary wall of tracheids fibres and vessels

Xeromorphic: refers to specialized structural features of plants adapted to live in dry habitats (xerophytes)

Xylem: the main tissue concerned with water conduction in vascular plants, characterized by the presence of tracheary elements

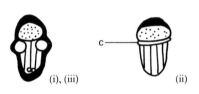

c

(i), (iii)   (ii)

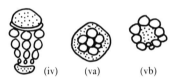

(iv)   (va)   (vb)

4 Histology of leaf, stem and root

Leaf, stem and root all contain one or more of the types of mechanical and transport systems described briefly in Chapter 2. Their tissues are composed of cells, most of which have been defined and illustrated in the glossary. Here we are going to examine a little of the diversity of organization of these cells into tissues. We shall see how the essential tissues are arranged in various organizational patterns in different families, genera and species. The development of the whole plant from the unspecialized meristematic cells will be better understood when the whole plant has been studied. That is why the usual textbook progression from cell–tissue–meristem–organs–whole plant is not followed here. Meristems are defined and described in Chapter 5.

A detailed account of xylem and phloem will be found in Chapter 6. Here, only the arrangement of these tissues is discussed.

The leaf

Leaves show a surprisingly wide range of forms when it is considered that in the majority of plants they perform two basic functions – the manufacture of food materials and the evaporation of water a process which provides power for the transpiration stream and aids cooling of the leaf in hot conditions.

Leaves can be classified in various ways, for example by their shape and size, their texture and colour, and the degree of hairiness, to name but a few. These variable features are frequently reflected in different internal tissue arrangements. Some modifications are typical of plants which grow under particular conditions, but other features may owe more to the genome than to the habitat. Ecological adaptations are discussed in Chapter 7, but have brief mention here also. Leaves differ from most stems and roots in that they are almost all primary organs, although some secondary growth can occur, as, for example, in the vascular supply of some gymnosperm leaves and in the leaf bases of some monocotyledons with secondary stem growth. However, large changes do not occur in the shape or thickness of dicotyledonous leaves after primary growth has ceased. Primary growth occurs at the basal (intercalary) meristem of many monocotyledonous leaves for a long time after maturity of the distal portions. Grass in lawns would not recover from cutting and continue to grow, if this were not so.

Some leaves are ephemeral, and are quickly shed, leaving nothing but a scar or perhaps a basal, membranaceous sheath as in *Elegia*. Then the stem takes over the functions of the leaf in such xeromorphs. Other plants shed their leaves at times of physiological drought, for example frozen ground, as is demonstrated in north and south temperate mesophytic trees and shrubs. The leaves on herbaceous perennials, biennials and annuals last for one growth season only. Some trees and shrubs of temperate regions (or at temperate altitudes in the Tropics) have reduced leaves, e.g. *Pinus*, *Cedrus*.

Many families have members with leaves which thrive for more than one season – and are thus termed 'evergreen'. Plants with long-lived leaves are not confined to particular climatic or altitudinal zones. The evergreens include such plants as the conifers mentioned above, as well as broad-leaved plants like *Camellia*, *Borassus*, *Phoenix*, *Rhododendron*, *Ilex*, some *Quercus* species, *Coffea* and *Ficus*.

It is interesting to note that some desert plants develop their leaves only after rain, e.g. *Schouwia* of the Cruciferae, and lose them in periods of drought, or like many bulbous plants, e.g. *Narcissus*, *Tulipa*, *Albuca* and plants with corms, e. g. *Gladiolus*, *Watsonia*, *Crocus*, have leaves which grow after the wet season and die down, leaving the underground storage organs safe from desiccation during the dry period.

Aquatic plants may have hibernation periods, where leaves die at the end of a season, e.g. *Potamogeton*, *Stratioles* and *Nymphaea*.

So it is apparent that there is no such thing as a 'typical' monocotyledous or dicotyledonous leaf. Except in extreme cases of reduction, as, for example, in some aquatic plants or xeromorphs, most plants have leaves which are made up of a combination of various essential components – the mechanical and supply systems, tissue in which photosynthesis is carried out, and the outer skin or epidermis.

General leaf anatomy

Figure 4.1 shows a leaf of *Ilex aquifolium* in T.S. and surface view. The epidermis is the boundary between the atmosphere and the other tissues. Its cells are specialized for this function. There is a thin cuticular covering. The outer walls of the epidermal cells are often thicker than the anticlinal and inner walls. In this dorsiventrally compressed leaf, the upper and lower surfaces are different, as can be seen from the figure. Stomata occur among the cells of the lower surface. The chlorenchyma, of which the mesophyll is largely composed, consists of palisade-like cells on the adaxial side and of more loosely-arranged spongy cells with larger intercellular spaces on the abaxial side. Part of the vascular system is shown – the large midrib bundle and the smaller secondary veins. All have phloem to the abaxial side and xylem to the adaxial side. Frequently the larger and many of the smaller bundles have a cap of sclerenchyma cells at the phloem pole.

Strengthening systems in the leaf

There may be additional marginal strands of sclerenchyma, and some fibre strands or girders associated with the vascular bundles (Fig. 4.2 *Aegilops crassa*, *Phalaris canariensis* and *Agave franzosinii*). Collenchyma is frequently present in the raised ribs above and below the midrib bundle, and is also occasionally found in similar positions in relation to the larger secondary veins.

The epidermis

Epidermal cells vary a great deal from species to species, particularly as seen in surface view. Many monocotyledons, and in particular those with strap-shaped or axially elongated leaves have elongated cells which are arranged in well defined longitudinal files. These cells may be 4–6 or more sided; their anticlinal walls

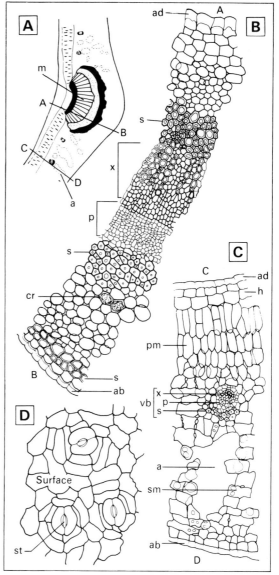

Fig. 4.1 *Ilex aquifolium* Leaf T.S. and surface. **A**, low power (× 22), diagram of midrib region, A–B and C–D indicate where detailed drawings **B** and **C** were taken. **B**, detail of midrib T.S. × 130. **C**, detail of lamina, T.S. × 130. **D**, abaxial surface × 200 a, air space; ab, abaxial epidermis with thick outer wall; ad, adaxial epidermis with thick outer wall; cr, crystal; h, hypodermis; m, midrib bundle; p, phloem; pm, palisade mesophyll; s, sclerenchyma; sm, spongy mesophyll; st, stoma; vb, vascular bundle; x, xylem.

may be straight, curved or sinuous. Sometimes the outlines of these walls are more sinuous near the cell outer wall than near the inner wall. Figure 4.3 shows a range of named examples of cell forms.

The epidermal cells of leaves of grasses fall into two distinct classes, described in the literature on

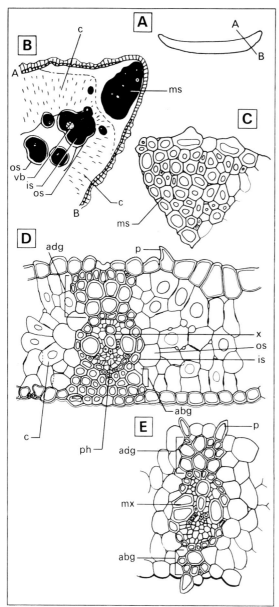

Fig. 4.2 Strengthening tissue in the leaf, as seen in T.S.
Sclerenchyma in *Agave franzosinii*, **A** and **B**; *Aegilops crassa*
C and **D**; *Phalaris canariensis*, **E**. A, outline of leaf T.S. to show
location of diagram **B**, ×40. **C**, leaf margin, ×109. **D**, ×109 and
E, ×230, vascular bundles and their associated bundle sheaths
and girders. adg, adaxial sclerenchyma girder; abg, abaxial
sclerenchyma girder; c, chlorenchyma; is, inner bundle sheath;
ms, marginal sclerenchyma; mx, metaxylem; os, outer bundle
sheath; p, prickle hair; ph, phloem; vb, vascular bundle;
x, xylem.

grass anatomy as 'long' and 'short'. These two size
classes should not be confused with variations in
cellular dimensions that are to be seen over veins

(costal cells) and between veins (intercostal cells).
The true 'short' cells are frequently suberized, or they
may contain silica bodies. Even small fragments of
leaf from a member of the Gramineae can often be
identified to the family level on the basis of the epidermal cell character.

The majority of dicotyledons and many monocotyledons without axially elongated leaves, e.g.
Smilax, *Gloriosa*, tend to have epidermal cells of
irregular shapes and sizes. They have straight, curved
or sinuous anticlinal walls. Because dicotyledon
leaves lack a basal meristem, but grow in area by regions of cell division (see p. 58) their epidermal cells
are rarely arranged in clear rows. Figure 4.4 shows a
range of cell types from named plants.

Anticlinal walls of both monocotyledons and
dicotyledons can be either very thin and hardly visible
from the surface or they may range through degrees of
thickness to very thick, so that the lumen of the cells
appears very small from the surface (Fig. 4.5). In
dicotyledons, as in monocotyledons, the costal cells
frequently differ from those of intercostal regions;
they tend to be elongated in the direction of the veins.

Sometimes the cells of the upper and lower surfaces of leaves may be similar, but more often they are
not alike. The dissimilarity may be in cell size and
wall thickness, or merely in the absence of stomata
from one surface.

Cells at the margins and the tip of the leaf are often
narrower than the rest, and have thicker walls. Some
marginal cells may develop into unicellular or multicellular prickles.

Measurements of epidermal cells have been made
to try and distinguish between closely related species.
If enough careful measurements are made and a sta-
tistical analysis carried out, significant differences
may be detected. Unfortunately, this seemingly valuable method is limited in usefulness by the natural
variation in size within different specimens of the
same species, or even amongst cells from different
leaves on the same plant. Sun and shade leaves, for
example, can differ in this respect.

Even if absolute size differences may seem to be
unreliable in many instances in distinguishing between species, the proportion of length to width of
epidermal cells can often give useful data for comparison. This length:width ratio can be fairly constant in
a species, even if the cell size varies phenotypically.
The importance of selecting leaves for comparison
from comparable positions on the various plants
under study cannot be overstressed. Normally, one
would select mature, vigorous leaves. The eye can
play tricks, and it is easy to be misled about length:
breadth ratios unless they are actually measured –

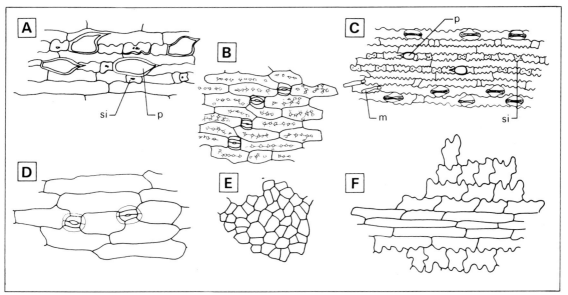

Fig. 4.3 Monocotyledonous leaf surfaces. A, *Phalaris canariensis*, × 240. B, *Kniphofia macowanii*, × 80, note cuticular pattern. C, *Arundo donax*, × 120, note microhairs. D, *Clintonia uniflora*, × 70. E, *Smilax hispida*, × 150. F, *Gloriosa superba*, × 54, note elongated costal cells over vein and cells with sinuous walls between veins (intercostal cells). m, microhair; p, prickle hair; si, silica body.

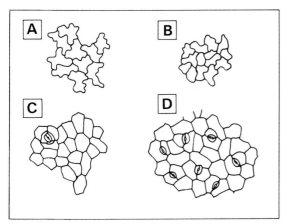

Fig. 4.4 Dicotyledonous leaf surfaces (abaxial). A, *Acacia alata*; B, *Aerva lanata*; C, *Plumbago zeylanicum*; D, *Cassia angustifolia*, all × 120.

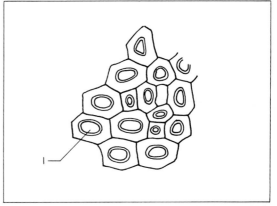

Fig. 4.5 *Gasteria retata*, leaf surface, × 145, showing very thick anticlinal cell walls and small lumina (l).

look at the diagrams in Fig. 4.6. Figure 4.6 also shows examples of epidermis of named plants in T.S.

Cuticular markings and patterns
The cuticle, and the outer part of the wall of the epidermis which it covers and grades into, is patterned in many plants. If the pattern is of low relief it will not show strongly in sections and may be faint in surface view. The strong pattern in *Aloe*, for example, may often be obscured by the granular appearance of the interface between cuticle and epidermis (Fig. 4.7).

Although many patterns can be seen with the light microscope, either on intact cells or with detached cuticles or surface replicas, the scanning electron microscope comes into its own in such surface studies.

In Aloes and Haworthias the range of cuticular/outer cell wall patterns is such that individual species or groups of species can probably be identified by their particular pattern. Striations are quite common, as are papillae. Some patterns are shown in Fig. 4.8.

Sometimes the cuticle and its markings are masked by a covering of wax. Most people are familiar with the waxy 'bloom' on apples and plums, and they have noticed that some leaves have a dull sheen on them,

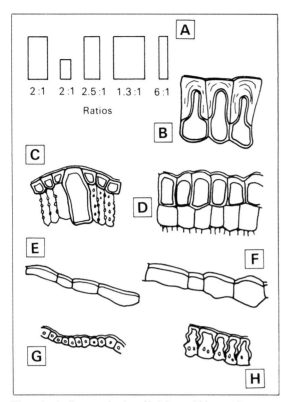

Fig. 4.6 A, diagram of ratios of height to width; note how difficult it is to judge these by eye. B–H, epidermis of selected plants in T.S., × 145. B, *Gasteria retata* note the thick outer wall and the outer part of the anticlinal walls. C, *Dielsia cygnorum*, note that some cells are larger than others. D, *Elegia parviflora*, note the double epidermis. E, *Cistus salviifolius*. F, *Gloriosa superba*. G, *Pinus ponderosa*, note the very thick walls. H, *Thamnochortus scabridus*, note the wavy anticlinal walls; pits are also visible.

for example, cabbage or *Agave*. Few realize that many other plants also have a waxy covering, since this may be very thin or easily removed. Many xeromorphic plants exhibit a waxy covering (see Chapter 7).

Wax takes on many crystalline forms, and may also be present as a melted-down layer. Some plants appear to go through a daily cycle in which wax crystals of one form melt, and recrystallize in another form. Many xeromorphic monocotyledons have a large proportion of their stomata plugged by wax. The SEM picture in Fig. 4.9 shows some wax flakes in *Aloe lateritia* var. *kitaliensis*.

Stomata

Stomata may be present on both surfaces, or the upper or lower surface only. They are characteristically absent from submerged aquatic leaves but are often present on the upper surface of floating leaves, e.g. *Nymphaea* and *Victoria*. In most gymnosperms and evergreen angiosperms, they tend to be on the lower surface only, but in some angiosperms with aerial leaves the distribution varies from species to species, depending to some extent on whether the plant is a xerophyte or a mesophyte.

The stomata may be superficial, that is, with the guard cells level with the surface of the leaf, or sunken, with a small outer chamber above the guard cells. Although many xerophytes have sunken stomata, and the majority of mesophytes superficial stomata, this is not invariably the rule. There may be particular adaptive advantages in each arrangement under certain circumstances; it may not be clear why some apparently 'unadapted' species survive while others around them are modified to a greater or lesser degree.

Of most interest to the taxonomist, or to the person wishing to identify a small leaf fragment, is the arrangement of subsidiary cells where these are present. Some of the various common types are illustrated in Fig. 4.10. Those stomata which lack subsidiary cells are called anomocytic. The cells surrounding each stoma are not recognizably different or distinct from the remaining cells in the mature epidermis. Such stomata occur in the Ranunculaceae, for example. Stomata with two subsidiary cells, one at either pole, are called diacytic; *Justicia* and *Dianthus* species have such stomata. Stomata with two subsidiary cells, one on either flank, are termed paracytic; these occur in for example, *Juncus*, *Sorghum*, *Carex* and *Convolvulus* species. The paracytic type also includes species with a number of subsidiary cells in a parallel arrangement on either flank. Tetracytic stomata, with four subsidiary cells, can be seen readily in *Tradescantia*; here one cell occurs on either flank and one at either pole. If there are three cells of unequal size surrounding the guard cell pair the stoma is called anisocytic, as in *Plumbago*. Cyclocytic stomata have a ring of subsidiary cells of more or less equal size and the individual cells are not very wide, whereas in the actinocytic type the subsidiary cells radiate strongly. Naturally enough, there can be intermediate forms which cannot easily be classified. Aberrant forms are also frequent – for example, two paracytic stomata may share one of the subsidiary cells.

Although occasional species exist which have several types of stomata on one leaf, most have one type only. This means that by noting the type of stoma present, the identity of a plant can be narrowed down – but, of course, many families share the more common paracytic and tetracytic types, so the combination of all characters available must be seen to fit

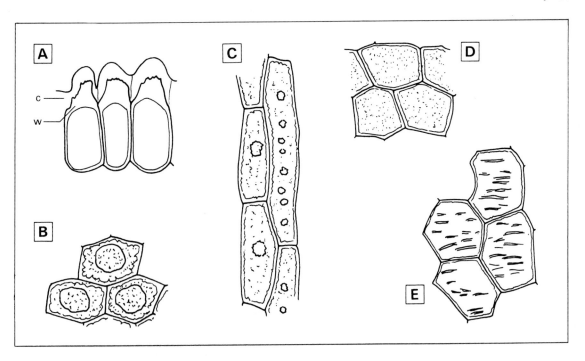

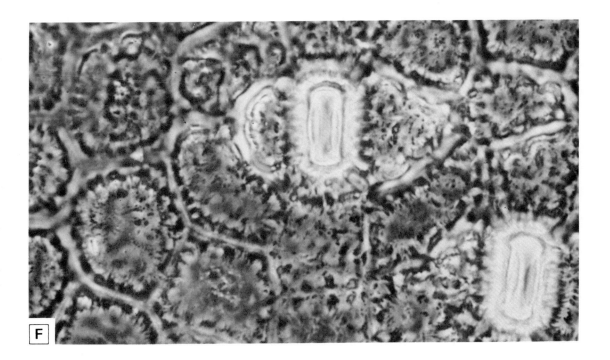

Fig. 4.7 Cuticular-cell wall patterns in *Aloe* leaves. A, *A. ferrox*, epidermal cells in T.S. showing the irregular interface between the cuticle, c, and the outer cell wall, w. B–E, leaf surfaces. B, *A. ferrox*. C, *A. cooperi*, note rows of papillae. D, *A. deserticola*, lack of prominent pattern. E, *A. cilliaris*, short, raised transverse ridges, all × 290. F, Anoptral contrast, *A. branddraaiensis*, × 590, the granular interface makes interpretation of the cuticular pattern difficult.

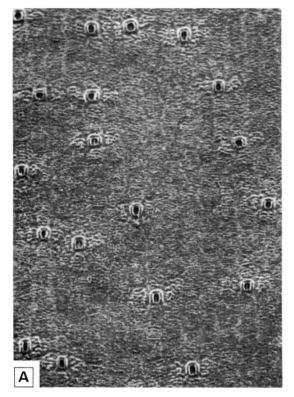

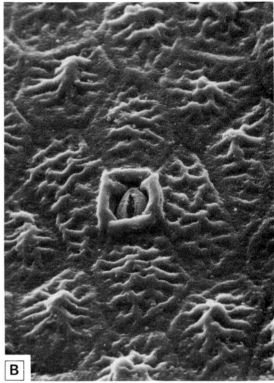

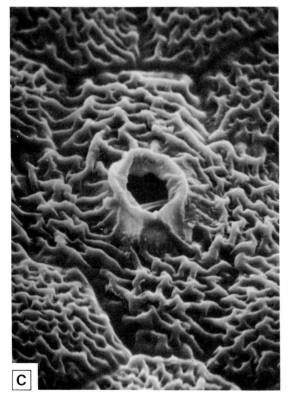

Fig. 4.8 Cuticular patterns are more easily seen using the scanning electron microscope. A, low power view ($\times$ 100) of *Aloe rauhi* $\times$ *A. dawei* showing distribution of stomata. B, *Gasteria lutzii* $\times$ *Aloe tenuior* var. *rubra*, $\times$ 1,250. C, *Haworthia cymbiformis*, $\times$ 1,250. Note that the rim to the stomatal pore is 4-lobed in the hybrid plants, an *Aloe* characteristic. The *Haworthia* belongs to the group of very succulent species within the genus, and has lobes which are fused into a cylindrical collar.

with reference material before an identification can be made.

There are other stomatal types, and indeed the ferns provide some interesting forms, the polocytic, with the guard cell pair towards one end of a single subsidiary cell and the mesocytic type, where the guard cell pair is in the centre of a subsidiary cell are two such examples.

It is all too easy to think that some forms of arrangement of subsidiary cells must be primitive, and some more advanced. By speculating, phylogenetic sequences can be postulated, and interrelationships suggested. One great danger in doing this arises because a mature stomatal type may be formed by more than one developmental sequence in different groups of plants. Perhaps we need two systems of naming stomatal types, one taking into account the mature form, and used only for identification, and the other derived

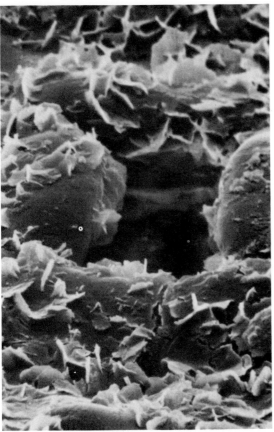

Fig. 4.9 *Aloe lateritia* var. *kitaliensis*, × 3,000; wax flakes on the four lobes surrounding a stoma. The guard cells are deeply sunken and can just be seen.

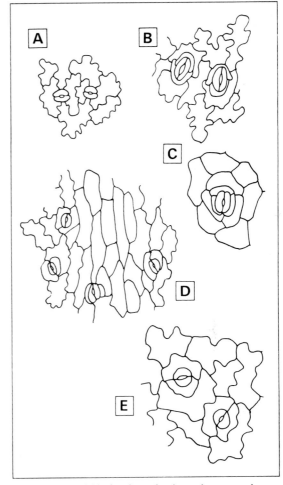

Fig. 4.10 Adaxial leaf surfaces, showing various stomatal types. A, *Chrysanthemum leucanthemum*, anomocytic, × 109. B, *Justicia cydonifolia*, diacytic, × 218. C, *Plumbago zeylanicum* anisocytic, × 218. D, *Convolvulus arvensis*, paracytic, × 109, note elongated cells over veins. E, *Acacia alata*, paracytic, × 218.

from a study of the ontogeny of the stomata and used by the phylogeneticist or taxonomist. Figure 4.11 shows two possible ways by which paracytic stomata may arise. In the first route the guard cell mother cell (meristemoid) divides first to produce two cells, then each of the flanking cells divides to form one subsidiary cell. The second pathway involves the division only of the guard cell mother cell. Either before or after division of this cell, two flanking cells divide to form one subsidiary cell each.

Sometimes mature stomata may appear at first sight to have no subsidiary cells. A study of the early stages of development could show that cells surrounding the guard cell mother cell divide in a particular way that differs from that normally occurring among the other epidermal cells. Many *Aloes* appear to have four subsidiary cells, whereas up to 8 cells surrounding stomata have anticlinal walls which are oblique. Most other cells in areas not adjacent to stomata have transverse walls. The oblique walls are the product of additional divisions in cells next to the guard cell mother cell.

Sometimes stomata are specialized to exude droplets of liquid water. They may simply be 'giant' stomata, larger than the others on the leaf, as in some members of Anacardiaceae. They might be specialized, and elevated at the end of a small mound situated at the termination of a small veinlet. Structures through which droplets of water may exude but which have non-functional guard cells are called hydathodes. Salt glands are a type of hydathode modified for the exudation of salt water. They are often surrounded by an encrustation of salt. Examples of hydathodes may be found in Saxifrages and salt glands in *Limonium* (Fig. 4.12).

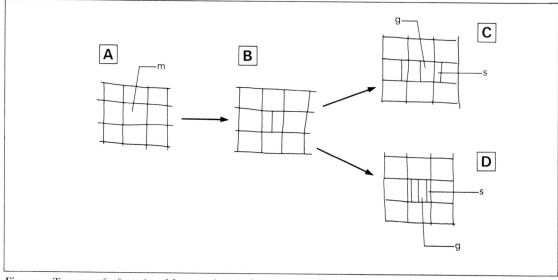

Fig. 4.11   Two routes for formation of the paracytic type of stoma. In A → B → C, guard cells are derived from the cells flanking the guard cell mother cell. In A → B → D, the guard cell mother cell divides to produce two cells, each of which divides once more. g, guard cell; m, guard cell mother cell; s, subsidiary cell.

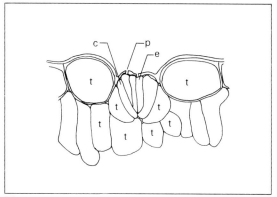

Fig. 4.12   *Limonum vulgare*, T.S. of salt gland from leaf, × 330. c, cup cell; e, excretory cell; p, pore; t, tannin-filled cells.

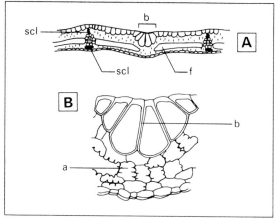

Fig. 4.13   *Pariana bicolor*, showing bulliform and fusoid cells. A, low power (× 54) diagram of leaf T.S., to show location of B, detail drawing, × 218. a, arm cells of mesophyll; b, bulliform cells; f, fusoid cell (typical of certain bamboos); scl, sclerenchyma girders.

Crystals and silica bodies occur in the epidermis but for convenience will be described in the 'mesophyll' section.

Mesophyll

Mesophyll can be taken to consist of the thin-walled parenchymatous cells containing chloroplasts, called chlorenchyma, and other thin-walled cells concerned with water, food or waste product storage. Also included are the specialized cells thought to be involved in leaf rolling, for example, 'motor cells' the bulliform cells (Fig. 4.13) of grass epidermis.

The classical division of mesophyll into palisade-like cells and spongy cells is misleading in its over-simplification. There are many intergrading cell shapes between the extremes. Figure 4.14 shows paradermal views of arm cells, part of the spongy tissue in *Clintonia*. Since some leaves lack a distinction of layers and others have very well-marked layers, the mesophyll can be used as an aid to identification. It cannot often be used as a guide to the taxonomic position of a plant, but within a group of related plants there may be close similarities of arrangement. Environmental variations will not alter arrangements which are rigidly controlled by the genome. For

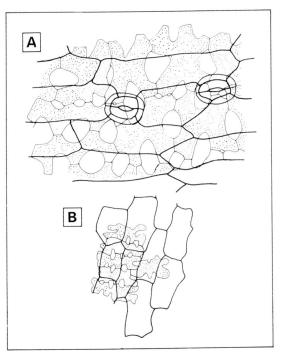

Fig. 4.14 *Clintonia uniflora*, paradermal views (through the epidermis) of arm cells, part of the spongy mesophyll (stippled). Note large air spaces between the cells. A, abaxial, × 115 B, adaxial, × 80.

example, palisade cells can be present adjacent to the upper or to the lower surface, or to both. There are, however, striking changes that can occur to the layers themselves. In some cases, the number of layers of palisade cells have been counted and this figure used as a diagnostic character. Since, in some plants, the leaves growing in the light may be thicker, and have more layers of palisade cells than those leaves which have developed in the shade, this is not a sound diagnostic character.

Pharmacognosists (who among other things, study plants and animals for natural products which might be applied in medicine) use a measurement called the 'palisade ratio'. This is particularly useful in defining small leaf fragments in powdered leaf products. This measure indicates the number of palisade cells that can be seen beneath an epidermal cell in surface view. An average figure is produced after many cells are counted. A statistically sound count will produce a fairly reliable typification of the material.

The arrangement of mesophyll cells may indicate whether a plant has the normal, C_4 photosynthetic pathway, or if a Kranz system operates. Radiating elongated cells around vascular bundles and parenchymatous bundle sheath cells with conspicuous chloroplasts are present in many plants which use a C_3, Kranz pathway in their photosynthetic cycle. There are, however, intermediate conditions in the anatomy.

In many gymnosperms and some angiosperms the mesophyll cells are plicate, with inwardly directed wall foldings (Fig. 4.15). The infoldings increase cell wall surface area and probably therefore make up, to some extent, for the smaller number of chlorenchyma cells which are often found in such leaves.

Specialized cells in the mesophyll can often help in making identifications. Firstly, there are those cells containing 'ergastic' substances. These are products related to the 'work' of the plant and may constitute stored food materials, such as starch, oil, protein and fat. They also include substances which cannot be related as yet to a particular function. If the function of a substance is not clear, it is often called a waste product. This is a rather lazy way out of the problem, particularly since many such substances are currently being identified as physiologically active by chemists. They often do not know which cells of the plant contain them and it could be that some of the so-called 'waste products' are really important to the plant. There is an obvious need for closer co-operation between morphologists and those extracting these interesting plant products.

Crystals, which are probably the best known of the ergastic substances, are very commonly thought of as waste products – again without sound evidence. They are usually composed of calcium oxalate and more rarely of calcium carbonate. Because they are of wide occurrence, they are of limited value to the applied anatomist. However, some families have never been recorded as having crystals, e.g. Juncaceae, the rush family. Others very frequently have a particular type, for example Liliaceae frequently have styloids (Fig. 4.16), so leaf fragments from the Liliaceae and Juncaceae families might be separated. However, other monocotyledonous families, such as the Iridaceae, have crystals similar to those of the Liliaceae and such diagnostic characters must be used carefully and always in conjunction with others.

In the dicotyledons a particular 'saddle-shaped' crystal is common in Leguminosae (Fig. 4.16) and its presence along with other features can help in distinguishing members of that family from others. Various prismatic and cluster crystals also shown in Fig. 4.16 have a very wide and scattered distribution through many families. Their presence in a fragment to be identified is only of real diagnostic value if they match exactly the type found in properly named reference material which has been narrowed down, using other characters, as probably being the species concerned.

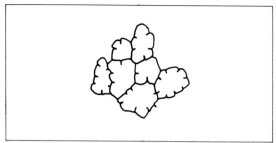

Fig. 4.15 *Pinus ponderosa*, plicate mesophyll cells from leaf T.S., × 145.

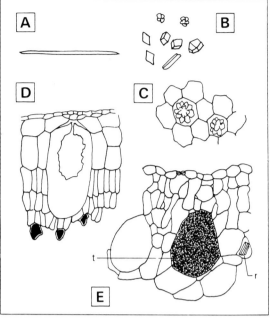

Fig. 4.16 Crystals, cystolith and tannin and latex cells, all × 125. A, styloid crystal, typical of many Liliaceae. B, *Acacia alta*, crystals from leaf. C, cluster crystals in *Passiflora foetida* leaf. D, *Ficus elastica*, leaf T.S. showing cystolith; dark cells contain latex. E, *Oscularia deltoides*, leaf T.S. with large tanniniferous idioblasts (t), and raphides, (r).

Crystals can be associated with particular tissues, for example, in a sheath round the veins, or they may occur in special idioblasts in the mesophyll. Sometimes there are no large crystals but merely a fine 'crystal sand' in the lumen of certain cells. Cystoliths are a special example of idioblasts; they occur in relatively few plants, e.g. *Ficus elastica*, and are illustrated in Fig. 4.16.

Silica bodies are similar in appearance to crystals. They normally occur in cells next to fibres or other lignified tissues, or in the epidermal cells, particularly those near to fibrous cells associated with bundle sheaths. However, since silica bodies are amorphous and not crystalline in structure, they can be distinguished from crystals by simple tests. They are in reality small opals! They do not show birefringence when viewed between crossed polars in the polarizing microscope, i.e. they do not shine brightly, as crystals do. In addition, they often turn pink when treated with a saturated solution of carbolic acid, and I know of no crystals which do that. **Be careful to keep the carbolic acid off your hands** if you try this test.

Many silica bodies occur in epidermal cells, usually one, but occasionally more to a cell. Since they are easy to see, it is worth examining a simple epidermal strip or scrape from one of the grasses, Cyperaceae – particularly *Carex* species – or a palm leaf surface, e.g. *Borassus* species. In *Bambusa vulgaris*, of the bamboos they are almost cuboid, as shown in Fig. 4.17. Those of *Zea* and *Agrostis* are also illustrated, together with some others from grasses or sedges which may be easily available to you.

There are silica bodies of many shapes and sizes in the grasses and extensive taxonomic use is made of them. Their form can be of help in the identification of fragments of cereal or grass which may have constituted part of the diet of an animal whose feeding habits are under investigation. They survive digestion and can be found in quite remarkable situations. For example it is present-day practice to use cow dung in the clay when bell founding, and it was thought that medieval bell-founders also used cattle dung to reinforce the clay of their bell moulds. Fragments of bell moulds from ruins of a thirteenth-century chapel at Cheddar were examined for such evidence – and there were leaf or chaff surface fragments together with silica bodies (Fig. 4.18) which had survived being eaten, fired in the clay by the molten bell metal and then several hundred years of burial!

The silica bodies of sedges are cone-shaped or conical, with flat bases. They often have small satellite cones around them as shown in Fig. 4.17. No grasses have this type of silica body.

Closely related families can sometimes be distinguished on the basis of presence or absence of silica bodies. For example amongst the Juncales, the rush family, Juncaceae and the Centrolepidaceae, which is a very small family of semi-aquatic plants from the southern hemisphere, lack silica bodies. On the other hand Resistionaceae, which are rush-like plants mainly from Australia and South Africa, typically have silica bodies shaped like small, spiky cannon balls. In Restionaceae the silica bodies rarely occur in epidermal cells, but more frequently in stegmata, specialized cells with thickened inner and anticlinal walls and thin outer walls. The thickening is often

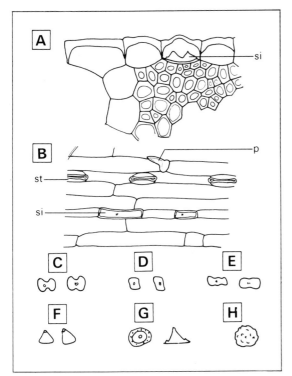

Fig. 4.17 Various silica bodies. A, *Cymophyllus fraseri* (Cyperaceae), leaf T.S., × 218; note locatio of silica bodies (si) in epidermal cells above sclerenchyma girder. B, *Aegilops crassa* (Gramineae), abaxial leaf surface, × 109, p, prickle hair; si, silica body; st, stoma. C–H, isolated silica bodies. C, *Zea mays* (Gramin.), × 200. D, *Bambusa vulgaris* (Gramin.), × 200. E, *Agrostis stolonifera* (Gramin.), × 200. F, *Evandra montana* (Cyp.), × 200. G, *Cyperus diffusus* (Cyp.), × 200, first body in surface view, second in side view. H, typical of many palms and Restionaceae, × 300.

Fig. 4.18 SEM photo (× 1,000) of a fragment of chaff from one of the Gramineae, found in fragments of a bell mould from ruins at Cheddar. Note the outline of silica bodies.

lignified and sometimes also suberized. But since most species of Restionaceae lack leaves, and the silica bodies occur in cells in the stem – this is probably not the place to be discussing them. However, the stem contains chlorenchyma and carried out leaf functions in that family.

The function of silica bodies is not understood. It is thought by some workers that plants cannot prevent the uptake of silicon with other elements, and that silicon in excess is deposited in an inert form – hence the proximity of silica bodies to veins. But this does not explain why many plants which must surely also take up silicon in excess do not also form silica bodies. Here is another unsolved mystery.

Tannins, polyphenolic substances usually characterized by their reaction with ferric chloride solution, when they turn blue-black, also have a scattered distribution through various plant families. Their chemical diversity is really a phytochemical problem.

The presence of tannins in special cells or cell layers can nevertheless be used as a diagnostic character even if their chemical identity is not known. A word of caution is due. Tannin may appear at certain seasons in some plants, so lack of tannins at a particular time of year is not a reliable feature and the plants cannot be assumed to lack them totally. Some tanniferous idioblasts are illustrated in Fig. 4.16. Some *Lithops* species owe their mottled brown appearance to tannin cells. One very familiar family rich in tannin is, of course, the Theaceae – to which the tea plant belongs.

The function of tannins is also little understood. They may act as an ultraviolet light shield in some plants where they are present in epidermal cells. Very strong sunlight can damage chloroplasts so such a screen could be useful. The astringent taste may protect leaves from being eaten.

Aleurone grains may be present and starch grains are described on pp. 91 and 94.

Sclereids can also occur as isolated cells in the mesophyll, or in well-defined positions in relation to other tissues, for example the vascular bundles. They range in form, as described in the glossary. Some of the types found, and the plants in which they occur are shown in Fig. 4.19.

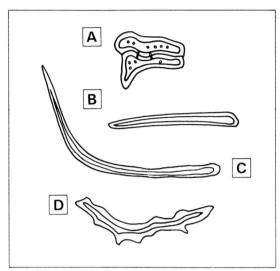

Fig. 4.19 Selected sclereids from leaves, all × 290. A, *Olivea radiata*. B, C *Olea europaea*. D, *Camellia japonica*.

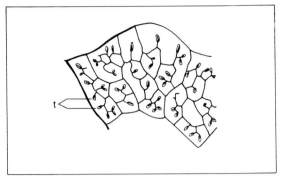

Fig. 4.20 *Plumbago zeylanicum*, paradermal view of veins to show open type of venation. Note enlarged tracheids (t) at veinlet ends, × 20.

Vascular tissues

The special cells whose function it is to conduct water and salts upwards from the roots and the cells which transport or translocate the substances synthesized in the leaf mesophyll and other tissues are grouped together in well defined strands called vascular bundles. In the leaf these are seen as the midrib and vein system. They are continuous, either directly or through the petiole, if developed, with the primary system of vascular tissue in the stem, or if secondary growth in thickness has occurred, with the secondary xylem and phloem.

Within the leaf blade, the vascular tissue is arranged in a pattern which appears to be under strong genetic control. The phenotypic expression of the genotype varies very little in overall pattern, although the number of bundles may vary in the leaves of plants of any one species found growing under a range of conditions. However, the main features which characterize a particular type of venation are constant enough for use in identification of fragments. It is rare for a family or genus to have a 'unique' pattern, but some families can be distinguished by the constancy of a particular type. For example, Melastomataceae have a vein running parallel with the margin in most species.

Vascular systems can be studied best in cleared material, subsequently stained with safranin. A range of main vein systems is readily observed. The angle at which the veins depart from the midrib can be a useful and relatively constant feature for a species. The nature of the vein endings of the final or smallest order

of branching is also used taxonomically. Figure 4.20 shows the 'open' type of vein ending in *Plumbago zeylanicum*.

The commonly held idea that all monocotyledons have parallel venation is quickly dispelled by examination of such leaves as *Bryonia* or *Smilax*. There are also dicotyledons which do not have the so-called net-like type of venation.

In most dicotyledonous leaves (except those which are greatly reduced and needle-like, e.g. *Hakea*) the phloem pole in a vascular bundle faces the lower (abaxial) leaf surface and the xylem pole the adaxial surface.

The monocotyledons are more diverse, and although many of them have the type of vascular bundle orientation described above, others have very different arrangements. Of course, the mature leaf develops from a meristem, and attempts to visualize the steps by which the more unusual leaf forms arise must stem from a study of the development itself and must not be inferred. Figure 4.21 shows the possible evolutionary pathways of several of these leaf forms, without suggesting that foldings and partial fusions, leading to total fusions actually occur during the growth of the present-day owners of a particular type.

To emphasize the danger of thinking of a sequence of folding and fusion processes, look at the vascular bundle arrangement of *Thurnia*, a plant from South America (Fig. 4.22). The relative position of the small vascular bundle system to the large bundle system, and the inverted orientation of these bundles would require a great deal of tortuous folding to achieve if it did not arise in one go, from a meristem!

Bundle sheaths

The phloem and xylem are not the only tissues present in the veins. They form the central core, around

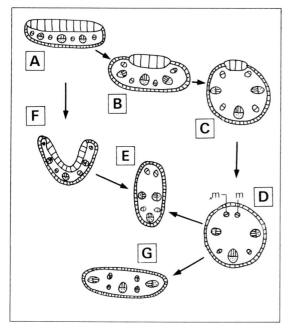

Fig. 4.21 Some possible evolutionary pathways leading to variations in vascular bundle arrangements in leaves. See text for fuller commentary. A, 1 row of bundles; adaxial and abaxial surfaces distinct. B, adaxial surface much reduced. C, smaller adaxial surface, leaf becoming cylindrical. D, loss of adaxial surface, leaf cylindrical, bundles in 1 ring, but 'marginal' bundles still distinct, (m). E, lateral compression, leaf of this type could arise from D or from F, where the adaxial surface is progressively lost. G could arise from secondary dorsiventral compression of the form in D.

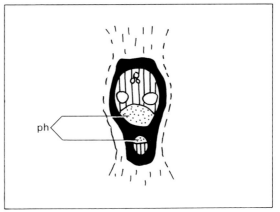

Fig. 4.22 Diagram of vascular bundle pair in *Thurnia sphaerocephala*, leaf T.S., × 57. The small bundle is inverted so that the phloem poles (ph) of the pair are opposite one another.

which sheaths of specialized cells are present. Two principal types of sheath exist, the sclerenchyma and the parenchyma sheaths. The sclerenchyma sheath is composed of fibres and/or sclereids. Sometimes the walls of these cells which face the phloem or xylem are more heavily thickened than the others. The parenchyma sheath is normally composed of much wider cells with thinner and usually relatively unlignified walls. If both types of sheath are present, the sclerenchyma sheath is often innermost. In some genera and species, three sheaths can be present, an inner, parenchymatous sheath, an intermediate, sclerenchymatous sheath and an outer parenchymatous sheath. This type occurs in *Frimbristylis*, Cyperaceae and indicates that the Kranz, C_3 metabolism is present. Named examples of the various types of sheath are illustrated in Fig. 4.23.

The sheaths may be complete, or present at the bundle poles as caps only, or present on the flanks of the bundles only. There may be adaxial or abaxial extensions of the sheaths reaching towards, or actually touching, either epidermis. The outline of

these girders as seen in T.S. can be used to distinguish species in some groups. In some plants, subepidermal strands of fibres may be present in line with vascular bundles. In a number of genera a hypodermis is present, composed of one or more layers of cells and occurring to the inside of the epidermis. The cells usually differ in shape or degree of wall thickening from both epidermis and mesophyll. It is a useful diagnostic feature. Its presence is often associated with plants adapted to grow in dry parts of the world.

Trichomes

Hairs and papillae are collectively called trichomes. Their occurrence and cellular structure are used extensively by taxonomists as an aid to identification, since there is such a wide range of form. When a plant possesses hairs or papillae, they are usually of a type or types characteristic of that species. It should be noted that various samples of a plant of a given species may range from being glabrous to very hirsute. This means that the numbers and density of hairs can be a poor character to use taxonomically except, perhaps, in defining subspecies or varieties, if there are other, linked characters to support the division. Also, although hairs are so diverse in form, there are very few of types which can be used even in the diagnosis of a family.

The greatest value of hairs is in identification. They are constant in a species when present, or show a constant range of form. Consequently, small fragments of leaf with hairs can often be matched with known material. If you look at the descriptions of powdered drugs of leaf origin in the British Pharmaceutical Codex you will see the hairs carefully defined.

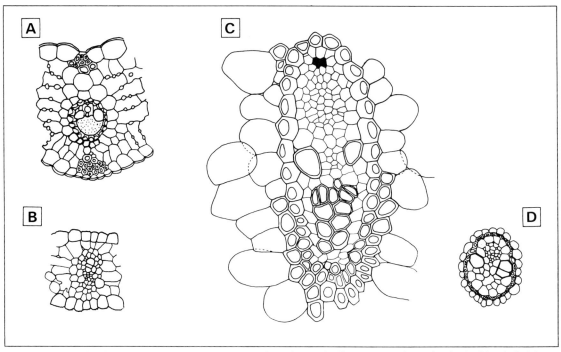

Fig. 4.23 Bundle sheaths. A, *Briza maxima*, inner mestome (sclerenchyma) sheath, outer parenchyma sheath, abaxial and adaxial sclerenchyma strands and radiate chlorenchyma, × 120. B, *Gloriosa superba*, × 120, parenchyma sheath only. C, *Cymophyllus fraseri*, well developed inner sclerenchyma sheath, outer, interrupted parenchyma sheath, × 120. D, *Fimbristylis*, three sheaths, inner parenchyma, followed by mestome sheath and outer parenchyma sheath, × 218.

In some families, individual species can be defined on the form of their hairs alone. Among these are Restionaceae and Centrolepidaceae which show good examples of simple, unbranched hairs (Fig. 4.24). The relative sizes of the basal cell and the cells of the free portion vary from species to species. The curious 'boat-hook' end of the cell in *Aphelia cyperoides* is diagnostic. *Gaimardia* has complex, branching filamentous hairs.

In the Restionaceae, *Leptocarpus* from Australia, New Zealand, Malaysia and South America has flattened, shield-shaped stem hairs. These are shown in Fig. 4.24. They are multi-cellular, diamond-shaped plates, held closely to the surface of the stem on short, sunken stalks. Until recently it was thought that *Leptocarpus* also occurred in South Africa, but the hair type and other internal histological differences show that the South African plants really belong to a distinct genus. A close relative to *Leptocarpus* in Australia is *Meeboldina*, the hairs of which have a diamond form, but have two large, thin-walled translucent central cells and a border of thick-walled cells with recurved micro-papillae which effectively zip adjacent hairs together so that they will strip off in a sheet if you try to pull one away.

At Kew, we have been sent samples of powdered leaf material from herbs which were thought to have foreign leaf added as an adulterant. Mint, *Mentha* species, was sent to the laboratory for purity tests on one occasion. It was found to contain *Corylus*, hazel leaf fragments. We also had an example of Marjoram being adulterated with *Cistus*. These impurities were readily spotted, because their hairs did not correspond to those of the authentic material (Fig. 4.25).

Some plants have hairs on both upper and lower surfaces but in many cases they are confined to the lower surface. Examples of leaves with hairs on both surfaces are to be found among the silver-leaved composites. The air in the hairs masks the chlorophyll in the leaf, giving the highly refractive, silvery or white appearance.

Hairs are divided into two major categories, the glandular and non-glandular (or covering) hairs. Glandular hairs (Fig. 4.26) include the stinging hairs of plants like the nettle, *Urtica*. Less familiar are the irritant hairs from the pods of *Mucuna*, the 'cow-itch' (Leguminosae), from the West Indies. These are mentioned here rather than in the chapter on flowers and fruits because although they illustrate a particular type not found on leaves, they help to

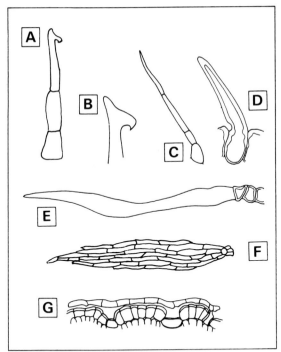

Fig. 4.24 Hairs in Centrolepidaceae, A–C and Restionaceae, D–G. A, B, *Aphelia cyperoides*, × 75 and 150. C, *Centrolepis exserta*, × 75. D, *Thamnochortus argenteus*, × 218. E, *Loxocarya pubescens*, × 218. F, G, *Leptocarpus tenax* surface view, × 113, longitudinal section, × 120.

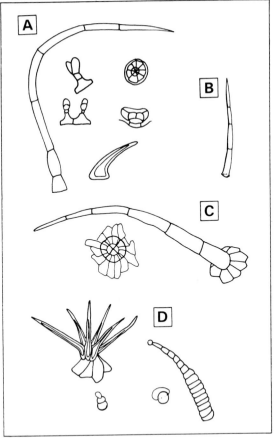

Fig. 4.25 A, *Mentha spicata*, range of hair types. B, *Corylus* hair (*Corylus* also has multicellular outgrowths). C, *Origanum vulgare* (marjoram) hair and sunken gland. D, *Cistus salviifolius*, range of hair types, one dendritic, the others glandular. All × 200.

broaden the concept of hair types. They are very sharp and brittle, and contain an oil which is irritant. We came across them being used by a landlord who wished to evict a tenant. He had sprinkled them liberally in the blankets of a bed, causing the tenant to come out in a rash! A sample of the fine powder composed of hairs was sent to Kew for identification.

Tobacco, *Nicotiana*, together with other members of the Solanaceae have rather characteristic glandular hairs. Some small cigars are enclosed in a paper made from macerated tobacco plant. In Great Britain, the law states that such cigars must be made entirely of tobacco. We once looked at some so-called tobacco papers to ensure that only *Nicotiana* had been used. The presence of glandular hairs of the correct type was quickly established, and epidermal cells with sinuous walls were also found. However, we also discovered some hardwood vessel elements and soft wood tracheids, and obviously other pulp had been added to strengthen the paper.

Simple glandular hairs are present on the leaves of plants which can trap and digest small insects and other small animals. Some of these are sticky, and some specialized to secrete digestive enzymes. *Pinguicula* and *Drosera* are examples.

Some of the fragrant (or less pleasant) essential oils occur in glandular hairs on leaves.

The non-glandular hairs are much more varied and diverse than the glandular. A range of types are illustrated in Fig. 4.27 and the plants on which they occur are identified in the caption. One or more of these plants or its relatives should grow near your home. The larger hairs will be visible with a hand lens.

As mentioned earlier, usually the hair type is only one of many characters used in an identification. However, some families are easily recognized by their hairs, e.g. the T-shaped hairs of Malpighiaceae (Fig. 4.27). Rhododendrons have been classified on the basis of leaf hairs, as an aid to the identification of species. Here, not only form, but also colour of hairs are used in the keys.

Micro-hairs, are very short, 2-celled hairs that are present on the leaves of some grasses, mostly from the tropics. Prickle hairs, which are usually prominent on

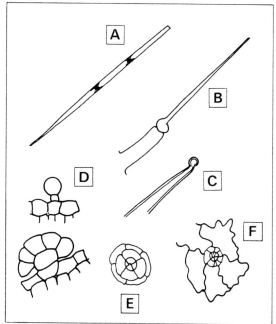

Fig. 4.26   Glandular hairs. A, *Mucuna*, brittle, sharp hair containing irritant oil droplets, × 145. B, C, *Urtica dioica*. B, low power (× 20), hair on multicellular base. C, fragile, sharp tip which can be broken off easily, × 290. D, *Salvia officinalis*, multicellular and bicellular hairs, × 290. E, *Justicia*, × 290. F, *Convolvulus*, × 145.

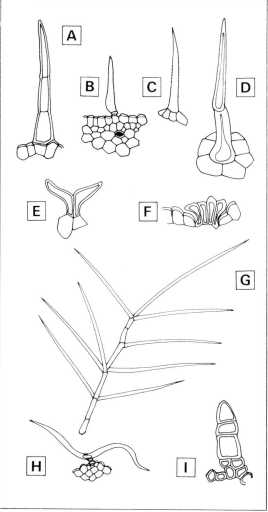

Fig. 4.27   Non-glandular hairs. A, *Salvia officinalis*, × 220. B, *Convolvulus floridus*, × 108. C, *Coldenia procumbeus*, × 220. D, *Justicia*, × 220. E, F, *Trigonobalanus verticillata*, × 220. G, *Verbascum bombiciforme*, × 54. H, I, *Artemesia vulgaris*, × 220 and 300.

margins and veins of grasses, are normally unicellular. They have thick walls which can be silicified.

A number of hairs are of commercial value. These are not leaf hairs, but usually come from the fruit or seed, e.g. *Bombax*, cotton and kapok.

The function of hairs is generally thought to be related to the water relations of a leaf. A densely hairy surface would tend to restrict the rate of flow of drying air.

The shape of the leaf lamina and the nature of its margin are outside the scope of this book. The student should bear in mind, though, that fragments of leaves, whilst not showing the shape of a leaf blade, may often have the characteristic dentations of the leaf margin and this can be a big help in identification.

Bulliform cells

Many leaves which are capable of rolling up in dry, unfavourable conditions, and re-open under conditions when there is no water stress, have special, thin-walled water-containing cells which enable them to make these movements. These are the bulliform or motor cells. Examples may be found in marram grass,

Ammophila arenaria and many members of the bamboosoid grasses. The shape, size and disposition of such cells can be used as an aid to classification and identification. Cells with similar properties are present at the pulvinus and also at the attachment regions of the leaflets to the rachis in many plants whose leaves fold at night.

Air spaces

These may be present in the mesophyll, between veins. These are much larger and usually more formal than the air cavities between cells of the spongy mesophyll, and often form by the breakdown of thin-walled parenchymatous cells between veins.

Water storage cells

Water storage cells are large, colourless and thin-walled, and usually lacking in conspicuous cell contents. Sometimes areas of the wall may be thickened in such cells. Water storage cells occur in many families, notably those which have representatives growing in arid conditions. Further details are given on p. 81 in Chapter 7.

The stem

The primary stem is considered here, together with the tissues making up the first stages of secondary thickening. Most plants, including annuals and even ephemerals, exhibit a degree of secondary thickening. The secondary xylem itself and the secondary phloem are described in Chapter 6.

It is easy to overlook the fact that at any time different parts of a plant are of different ages, and show varying degrees of secondary development.

Primary stems are equipped with an epidermis, often very similar to that of the leaf of the same species. This is followed to the inner side by cortical tissues, the outer layers of which (and the epidermal cells) may contain chloroplasts. Some cells acting as a physiological boundary between the cortex and stele are often present, forming a cylinder. They may be morphologically distinct as an endodermis, but sometimes they cannot be discerned as a separate layer.

Strengthening tissues can be present in the cortex or around the periphery of the stele (usually in the phloem), or in both positions. These tissues are usually in the form of axially arranged, rod-like groups of cells, with gaps between them. Only in stems with very limited growth in thickness do they form a complete cylinder, and this only when primary and secondary growth have ceased.

The vascular bundles can take up a variety of arrangements. In dicotyledons they usually occupy one ring, just to the inner side of the cortex. In monocotyledons they may form one ring, or may appear to be scattered in several to many rings, or lie without apparent order in the central ground tissue. The possession of several rings of vascular bundles is not the prerogative of monocotyledons. Several families of dicotyledons have this type of arrangement, notably those with climbing members, and also Piperaceae.

When vascular bundles are not scattered, there is frequently a parenchymatous centre to the stem. This may contain some sclereids or parenchyma cells with thickened (lignified) walls. In some plants the central parenchyma breaks down to form a canal. Diaphragms of specialized stellate cells may traverse such canals, and in some plants axially arranged diaphragms are also present.

Most dicotyledonous stems have nodes, where leaves are attached, with leaf gaps in the vascular system which become apparent when some secondary thickening occurs. The leaves usually each have a bud in the axil. If this develops, a branch gap is also formed. The internodes do not normally bear buds, unless these arise adventitiously. Monocotyledons have a range of types of shoot organization. They may have nodes where leaves are attached, as in grasses and sedges, or no formal node may be discernible in the internal structure, although the leaves appear to be attached to the stem in a similar way to those of the nodal plants when viewed from the exterior, e.g. in Palms. Since there is no cambium developed between the individual vascular bundles in monocotyledons, no leaf or branch gaps form. Because anatomy and gross morphology are so closely related, it is important to study the morphology of a plant as a whole organism before cutting it up to look at the cells and tissues. Indeed, for intensive and comprehensive studies, development should also be followed. Only if you examine morphology and development is it possible to be sure that you can locate similar parts of various species for comparative anatomical studies.

Cross-sectional appearance

The *cross-section* of a primary stem may have a more or less circular outline. However, it can take on one of a wide range of forms, some of which assist in the identification of a family – as in Labiateae, where the section is square – or may help to distinguish genera – for example many *Carex* species have stems with a triangular section. Often the outline is modified near to nodes or in regions of leaf insertion. Sometimes a wing or ridge of tissue in line with a petiole may continue down the internode as in, for example, *Lathyrus*. In general, the outline of the section taken in the middle of an internode would be described for comparative purposes.

Many stems have all or most of the following tissues, working from the outside inwards: epidermis, hypodermis, cortex (with both collenchyma and chlorenchyma, or either), an endodermoid layer (or a well-defined starch sheath), vascular bundles in one or more rings, or apparently scattered, and a central ground tissue or pith.

Sometimes a *pericycle* can be distinguished, but this is normally regarded as part of the phloem, and, rarely, a true endodermis, with casparian strips is present.

Epidermis

The *epidermis* can be one or more layered. The cells may be similar in form to those of the leaf of the same species, but more often they show more elongation in the direction parallel with the stem axis, and their anticlinal walls as seen in surface view are often not markedly sinuous. The outer wall is usually thicker than the anticlinal or inner walls. The proportion of cell length to cell width, or height to width as seen in T.S., can be used with caution as a diagnostic feature, but actual measurements are not normally sufficiently constant to be used in identification or classification.

A range of the normal measurements of a cell, with an average figure should be quoted in diagnostic descriptions where a sufficiently large sample has been examined to obtain sound data.

Stomata

In taxa with well developed leaves, the stem *stomata* tend to be much more sparse, but of the same type as in the leaf of the same species. When the stem is a principal photosynthetic organ, either supplementing or replacing leaves, stomata tend to occur in a higher frequency. Often the guard cells are aligned parallel to the long axis of the stem.

Trichomes

Hairs and *papillae* exhibit the same wide range as on the leaf, and examples are shown in Figs. 4.24–4.27. The type of hair can be of diagnostic value at species level, sometimes also at genus level, but rarely at family level.

Silica bodies

Silica bodies are often present in the stem epidermis or in other parts of the stem in species which have them in the leaves. There are leafless species of some families which have silica bodies in the stems. In the epidermis, the cells most likely to have silica bodies are those above fibre strands or girders. The range of form is shown in Fig. 4.17.

Cortex

The *Cortex* can be very narrow, and composed of few cell layers, or wide and multi-layered. The cortical zone is traditionally regarded as extending from epidermis to an inner boundary inside which vascular bundles are present. The inner boundary is often very indistinct, and sometimes vascular bundles from leaf traces may be present amongst cells which clearly belong to the cortex itself. Again we have an example of a tissue which is difficult to define clearly.

In a number of taxa, a *hypodermis* is regularly developed, and is quite distinct, e.g. *Salvadora persica*, Fig. 4.28. In other taxa a hypodermis is never found. However, because of the sporadic occurrence of the hypodermis in the taxa of vascular plants, presence or absence is of little taxonomic and small diagnostic value (except at the species level). Chloroplasts may be present in the collenchyma cells of the outer cortex, or in more or less well-defined layers of parenchymatous cells, as well as in palisade cells or cells of various shapes. Stems of leafless plants, e.g. some *Juncus* species often have a very formal and regular *chlorenchyma* arrangement. There are a few herbaceous plants which lack chloroplasts in cortical tissues, and these plants normally figure amongst those with abnormal modes of nutrition, e.g. *Orobanche*.

In species with a wide cortex the cells of inner layers are normally larger than those of outer layers and have few, if any, chloroplasts. In aquatic plants, large, formal air spaces may be present in the cortex which itself merges with central parenchymatous tissues.

Fibres and *sclereids* are a prominent feature in the cortex of many species. Often the grouping of fibres into strands with well-defined cross-sectional outlines and in characteristic positions in the cortex will help in the identification of a plant. Fibres can show individual peculiarities which enable one to identify even isolated strands. This is particularly true for fibres which are of economic importance, many of which are cortical in origin, e.g. *Linum*, flax and *Boehmeria*.

Crystals often occur in cells of the cortex and the central ground tissue. The cluster crystals or druses are probably the most common type, but solitary crystals of various shapes and sizes, similar to the range exhibited in leaves are of widespread occurrence. Raphides are not of widespread occurrence in dicotyledons.

Endodermis

In some plants the inner boundary of the cortex is

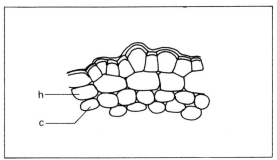

Fig. 4.28 Hypodermis in stem of *Salvadora persica*, × 290. c, cortex; h, hypodermis.

well differentiated into an *endodermis* and shows casparian strips, e.g. Hydrocharitaceae. In *Helianthus* the cells are well defined and are rich in stored starch, but lack casparian strips; they constitute a 'starch sheath'. In many other plants where the cells in this zone are morphologically distinctive but lack casparian strips or stored starch, they are probably best called an '*endodermoid layer*'. Some people prefer to use the term endodermis for this cell layer even when casparian strips cannot be seen.

Vascular and strengthening tissue

Many monocotyledons and some dicotyledons have a well-developed cylinder of sclerenchyma to the inner side of the endodermoid layer. In it are embedded some of the vascular bundles, often all the small bundles, and sometimes some of the larger bundles as well. Most dicotyledons lack such a cylinder. Their 'open' vascular bundles, each with a cambium, are arranged in a ring. Each bundle may have a cap of fibres to its outer side, but the flanks of the bundles are not enclosed, so enabling unhindered development of the interfascicular cambium to produce a continuous cambial ring in secondary growth.

In many of those monocotyledons lacking a sclerenchymatous cylinder, individual bundles have a sheath of sclerenchyma. This is frequently only a few layers thick on the flanks, and several layers thick at the xylem and phloem poles. An outer parenchymatous sheath features in a number of these plants. Since there is no cambium as such, the enclosing of vascular tissues in these plants does not affect normal growth and development. In the monocotyledons which have a vascular plexus at the nodes, the sclerenchyma sheaths become discontinuous. If there is an intercalary meristem, the bundle sheaths are poorly developed in the meristematic zone.

In monocotyledons stems the vascular bundles are either collateral, with one xylem and one phloem pole, or amphivasal, with the xylem encircling the phloem. Amphivasal bundles are frequent in rhizomes and less frequent in inflorescence axes. Figure 4.29 shows a range of bundle types. Dicotyledons usually have open bundles, but climbers, e.g. Cucurbitaceae may not develop interfasicular cambium, and remain as discrete strands. Since such bundles are present in a compressible parenchymatous matrix, twisting and distorting of such stems during the process of climbing does little damage to the vascular bundles themselves. In cucurbits and a number of other climbers the phloem is particularly well developed. It is present in two strands, one on either side of the xylem on a radial axis as seen in T.S. Bundles of this kind are called bicollateral. Because they occur in relatively few families their presence in a sample to be named is most helpful in reducing the field for further analysis.

Vascular bundles which have phloem surrounding the xylem are termed amphicribal.

There are anomalous plants in which individual bundles have cambium and extend radially in secondary growth, without uniting laterally, as for example in the Piperaceae (Fig. 4.29). Several other abnormal forms of bundle arrangement are found among the dicotyledons.

It is a dangerous practice to attempt to define various types of vascular bundles on the observations made from a few transverse sections of monocotyledon stems. This is because over its length, a single bundle can exhibit changes in its cross-sectional appearance. A newly-entering leaf trace can look different from a main axial bundle – but the two are parts of the same strand, the first being the increment before bridging and connecting with the strands from other parts of the plant.

The course of vascular systems in monocotyledon stems has been studied for many years and is under active investigation at present. Cine films produced by photographing serial sections have enabled researchers to understand for the first time the true complexity of such stems as those of palms and Pandanaceae. Most attempts at drawing from serial sections or dissections previous to the use of the cine method gave misleading results. A new area of comparative anatomy is emerging – the study of whole vascular systems. The results of this study might well show basic types which underly the major phylogenetic divisions in the plant kingdom.

Within the vascular bundles, the phloem and xylem of primary systems show only axial systems of cells. The rays are a feature of secondary development.

Phloem in gymnosperms has well developed sieve

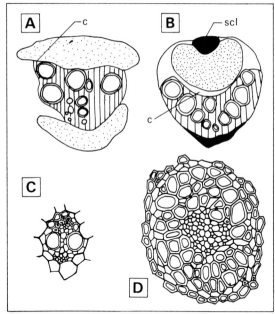

Fig. 4.29 Vascular bundle types from stems. A, *Cucurbita pepo*, diagram of bicollateral bundle, × 15. B, *Piper nigrum*, diagram of collateral bundle; cambium remains fascicular, × 15. C, *Chondropetalum marlothii*, detailed drawing of collateral bundle, lacking cambium, × 110. D, *Juncus acutus* detailed drawing of amphivasal bundle, × 220. c, cambium; scl, sclerenchyma.

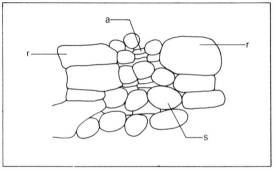

Fig. 4.30 Albuminous cells in gymnosperm phloem, *Acmopyle pancheri*, T.S., × 290. a, albuminous cell; r, ray; s, sieve cell.

tube elements with sieve areas or plates, and albuminous cells. In angiosperms the albuminous cells are replaced by companion cells (Fig. 4.30). It is thought by those who study phloem that an evolutionary sequence can be observed, from systems in which companion cells are poorly defined and the sieve tube elements communicate by rather scattered sieve areas on oblique walls to the most advanced in which sieve plates are very well defined and constitute the transverse end wall between elements in a sieve tube, and in which the companion cells are very well developed. Since the advanced sieve tube element has no nucleus, the organization of the element is carried out by the nucleate companion cell adjacent to it. Damage to the companion cell in this system may bring about failure of the element which it directs. Phloem is not the easiest of tissues to study with the light microscope and consequently it is only in recent years that the brilliant comparative studies such as those carried out by Professor K. Esau have made an impact on the botanical world.

Sieve plates can be seen easily in a number of plants for example, in the cucurbitaceae.

The first formed phloem is termed protophloem. It is often functional for a short period, and becomes compressed by the later-developing metaphloem.

Primary phloem can contain sclereids and fibres. Their presence helps in identification of fragments. Primary xylem is composed of protoxylem, in which the tracheary elements usually have spiral (helical) or annular wall thickenings. In metaxylem the wall thickenings can be more extensive, and breached by pits (with membranes) arranged in scalariform, alternate reticulate or less regular ways. The protoxylem has to be capable of considerable extension without breaking during the first phases of primary growth in length of the stem. However, the elements often do rupture, leaving a protoxylem canal. The more rigid metaxylem matures after this extension phase and in consequence is less liable to damage. Its structure does not have to allow for extension.

It is often difficult to decide if the protoxylem and those metaxylem fractions with annular or spiral thickenings are tracheids or vessel elements, since perforation plates can be very obscure and may even appear to be present in damaged, macerated preparations when they are, in fact, absent. Even with cells which have scalariform, alternate or reticulate pitting it may be hard to be certain if they are perforated or not, since perforations may be very small. This is of some importance, however. It is now widely believed that the narrow elongated imperforate tracheid is ancestral to the wider, shorter perforate vessel element. Therefore, plants which are vessel-less are thought to have primitive wood. The hinge of many phylogenetic systems swings on this delicate frame. Several methods are employed to try and determine if a cell is perforate or not. The tissues are usually macerated to separate the individual cells. These can then be examined with phase contrast, when intact pit membranes show up well. Another method involves flooding the macerate with indian ink, on a microscope slide. A cover slip is placed over the cells and gently tapped. The ink is then replaced by 50 per cent glycerine, drawn under the coverslip using filter paper from the opposite side. The indian ink contains solid particles, and if the cells are perforate, particles should be visible inside the lumina of the vessel elements.

More recently, macerated cells have been examined using the SEM, where membranes are readily observed. As with primary phloem, no radial system of cells is present in primary xylem, but fibres or sclereids and occasionally, parenchyma cells can be present.

Central ground tissue

The *central ground tissue* or pith is composed of cells which are usually parenchymatous, with simple, rounded pits in their walls. The walls may be thin, and composed largely of cellulose, or thickened with lignin. This matrix of cells can in various plants contain sclereids, tannin cells or crystalliferous cells, or combinations of the three.

Certain groups of plants have special cells or tubes containing latex. *Landolphia* has latex cells; most Euphorbiaceae have latex tubes. In instances where members of the Cactaceae and Euphoribiaceae have evolved to appear similar externally, it is very easy to distinguish the two anatomically. These plants of dry areas have only to be cut – latex will ooze from all the euphorbiaceous representatives and very few of the Cactaceae. Incidentally, **this latex can be poisonous and irritating to the skin, and even lethal.** Many members of the Compositae contain latex and *Taraxacum* was even grown experimentally in search of a *Hevea* substitute for rubber latex during the Second World War. *Hevea brasiliensis* itself is probably the best known latex producer. Amongst the monocotyledons, many members of the Alismatales contain latex canals with a secreting epithelial layer.

Latex tubes are among the longest cells or coenocytes; they often go on growing during the whole life of a plant.

Because the incidence of latex forming cells or tissues is restricted in the angiosperms, and further, because there are various types of cell, tube or canal, and since the laticifers can be articulated or unicellular their presence in a plant can be a great help in identification. Also, the distribution of these cells or tissues – in cortex, phloem, xylem or ground tissue – can be diagnostic.

The centre of the pith may be composed of parenchymatous cells, or it may be hollow, either as a single tube, or variously divided by transverse or longitudinal septa. The cells in the sepa or diaphragms can be of a simple, more or less dodiametric type, or they may be stellate or armed or branched

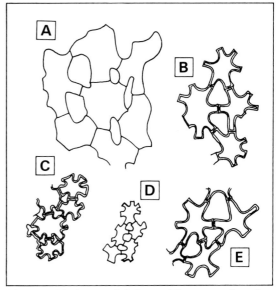

Fig. 4.31 Diaphragm cells in leaves of Cyperaceae, all × 218. A, *Becquerelia cymosa*. B, *Mapania wallichii*. C, *Chorisandra enodis*. D, *Mapaniopsis effusa*. E, *Scirpodendron chaeri*.

cells of various descriptions. A range of such cells (from leaves) is shown in Fig. 4.31.

The root

Primary roots have not been the subject of such full study as stems or leaves. They do, however, show a wide range of variation which is influenced both by environment, in terms of ecological adaptation, as well as by the genotype. Compared with stems and leaves, root fragments can be difficult to identify in the primary state. This is not entirely because they are relatively undescribed or poorly represented in reference microscope slide collections, but partly since there is, overall, less variation.

It will be recalled from Chapter 1 that the root is an organ which has to take strains or pulling forces. It rarely has to bend or flex, since it is usually found in a more or less solid medium. In consequence of this, the main strengthening tissues are positioned in the central region of the root and function like a rope.

Epidermis

In all except aerial roots and the non-anchored roots of aquatic plants, *root hairs* are usually present a short distance from the growing apex. These develop from the rhizodermis or root epidermis. Often the hairs arise centrally from the basal part of the cell; occasionally they arise from near one end. Again, whilst many root hair bases are level with other cells in the rhizodermis, in other plants they may be bulbous and protrude; they can be sunken into the outer cortical tissues (e.g. *Stratiotes*).

A short distance further away from the apex, the root hairs often die and shrivel, but in some plants the root hairs persist for a long time.

To the inner side of the rhizodermis an exodermis may develop. This characteristically consists of angular cells with somewhat thickened, lignified walls.

Cortex

The *cortex* is sufficiently variable to be used to assist in identification. Unfortunately from that point of view, the various types of cell arrangement seem to have more ecological than systematic significance.

Two basic types of cortex are recognizable amongst other, less frequent, variations. These are the 'solid' and the 'lacunate' cortex, Fig. 4.32. The 'solid' cortex is composed of parenchymatous cells which are relatively compact, with intercellular spaces confined to the angles between cells. It is usual for there to be a gradual increase in the size of such cells from the outer to the inner layers, but the innermost few layers are frequently of a smaller and more compact type of cell. Such an arrangement is common in both monocotyledon and dicotyledon roots on plants which grow in well-drained soil types. The 'lacunate' cortex has a few outer layers of compactly arranged cells and the innermost layers may be similarly compact. Between these, radiating plates of cells are seen in T.S., with large air spces between them, Fig. 4.33. In T.L.S. these may be seen as longitudinal plates, but more commonly, they are arranged in a net-like pattern, thus enclosing the air cavities or lacunae. Diaphragms of stellate and other cell types may traverse the lacunae, as in the stem. Most plants with this type of root cortex have their roots in ground which is periodically water-logged, or even immersed in water.

The cortex in aerial roots can be developed to extreme proportions and be many-layered. Frequently cells in this situation have specialized spiral or reticulate or irregular thickening bands and are capable of storing water absorbed from a humid atmosphere. Such a *velamen* is found, for example, in the aerial roots of epiphytic orchids, Fig. 4.32, and aroids.

The number of layers of cells in a cortex can vary in specimens of a given species, but within bounds. It

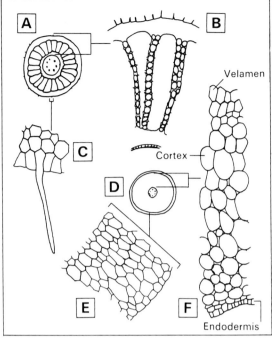

Fig. 4.32 Roots in T.S. A–C *Juncus acutiflorus*, A, diagram; B, lacunate cortex, ×54; C, root hair, ×218. D–F *Cattleya granulosa*, D, diagram; E, velamen, ×68; F, 'solid' cortex ×68.

might be possible to distinguish between species in a genus if some of them have many layers and others few, but this is not a very sound exercise.

Sclereids, fibres, tannin cells, mucilage cells and crystal containing cells can be found scattered in the parenchymatous cortical tissue in a wide range of families. Their presence and distribution can be helpful in narrowing down possibilities for identification, but are rarely of taxonomic significance.

Endodermis

The cortex on its inner side abuts onto the *endodermis*. This characteristic, physiologically active tissue is frequently one layer thick but in some plants it can be two or more layered. Although the endodermis can be composed of cells with evenly thickened walls, in the majority of plants the inner and anticlinal walls are more heavily thickened with lignins and suberins than the outer periclinal walls. Consequently, in T.S. they are readily distinguished from adjacent cell layers, the so-called U-shaped thickenings making them conspicuous (Fig. 4.34). At intervals, certain cells of the endodermis are thin-walled. These are the *passage cells* which are usually opposite protoxylem poles and are supposed to afford a more ready pathway

Fig. 4.33 *Stratiotes*, part of root T.S., SEM photograph, note air spaces in cortex, × 75.

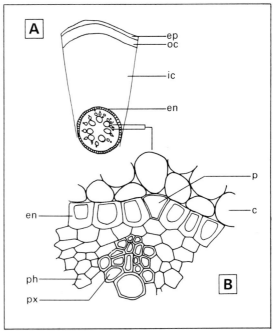

Fig. 4.34 Root endodermis *Iris* sp. A, low power, sector of root T.S., × 20. B, detail from A, × 290. c, cortex; en, endodermis; ep, epidermis; oc, outer cortex; p, passage cell; ph, phloem; px, protoxylem.

from root hairs, via the cortex to the protoxylem elements, for water and dissolved solutes. The remaining endodermis cells are supposed to restrict water flow between cortex and stele. The anticlinal walls of all endodermal cells are equipped with special suberized 'waterproof' impregnations, the *casparian strips* or bands. These are most easily seen in young, unthickened cells near to the root apex; they stain readily in Sudan III or IV. Because of the range of variation in cell height and width and differences in the shape and degree of wall thickening of endodermis cells throughout the monocotyledons and dicotyledons, it is often possible to give a close description of an endodermis which is characteristic for a species or group of species. There may be many species which fit any one description, but if authenticated material is available for comparative purposes in making identifications, then a close matching of endodermis cell types must be achieved for accurate identification. As with all other minute characters, the appearance of the endodermis could never be used on its own to identify an unknown plant, but if it is simply a matter of choosing between several possible plants, then a close match on the endodermis would be quite sound evidence on which to base the identification.

Pericycle

The cells of the next layers to the inside are usually smaller than those of the endodermis, and they frequently have relatively thin walls. They constitute the *pericycle*. Very few species lack a pericycle, among them members of the southern hemisphere Centrolepidaceae. Roots lack any nodes, and lateral roots arise endogenously, that is their growing points or apices first develop in the pericycle. Division of cells in this region produces a lateral root which has to grow through the tissues of the endodermis and cortex to reach the exterior of the primary root. Since the pericycle bounds the vascular system of the root, vascular continuity between the new lateral root and the main root can soon be established once active growth has started. Some roots may have quiescent, potential laterals in the pericycle which require some hormonal stimulation or the removal of a hormonal restraint before they develop. The relatively simple nature of the pericycle and comparative lack of variation from species to species renders it of little use as a diagnostic feature.

Vascular system

The vascular system can take on one of several forms

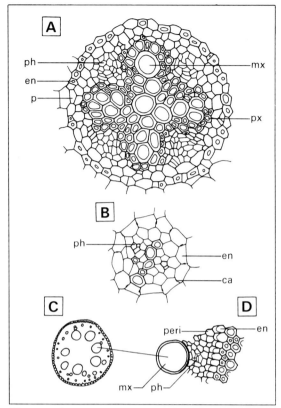

Fig. 4.35 Some root vascular systems, all but C, × 300, C, × 35. A, *Ranunculus acris* tetrarch root. B, *Echinodorus cordifolius* diarch root. C, D, *Juncus acutiflorus* polyarch root. ca, casparian strip; en, endodermis; mx, metaxylem; p, passage cell; peri, pericycle; ph, phloem.

(Fig. 4.35). In most dicotyledons there are between two and up to about six protoxylem strands alternating with phloem strands. Large numbers of species have triarch (3-stranded) or tetrarch (4-stranded) arrangements, or a mixture of both. If above 6 strands are present, the roots are described as polyarch. Most monocotyledonous roots are polyarch. The metaxylem tracheids or vessel elements are normally conspicuous, and are on the same radial axis as the protoxylem poles. Commonly in monocotyledons there is one ring only, but there may be more, or scattered metaxylem elements in the root centre. In most dicotyledons several metaxylem elements are grouped together in strands.

Although phloem is normally confined to the outer ring, occasional genera, e.g. *Cannomois* in Restionaceae, can have additional strands associated with the dispersed metaxylem elements.

The vessel element is thought to have had its origin in roots; in the least advanced plants, if vessels are developed at all, they are found in the root only,

and not in the stem or the leaf. The next stage of advancement is for vessels to occur in the root and the stem, and in the most advanced plants vessels occur in the root, the stem and the leaf. A number of plants have shorter, wider vessel elements in the root than in stem or leaf, bearing out the theory of the evolutionary sequence from the primitive narrow long elements to the more advanced shorter, wider elements. In supposedly primitive plants, then, it becomes of great interest to see if vessels are present in the roots. The methods used are as described on p. 50.

Vessel and tracheid wall pittings and thickenings are similar to those of the stem wood. The phloem cells of the root also have the same range of forms as are found in the stem.

The centre of the root may be made up entirely of xylem in the dicotyledon, or in monocotyledons or dicotyledons, it may contain a ground tissue composed of parenchyma with thin or thickened walls, sometimes termed pith. Sclereids could be present.

If a root is to be identified, it must be matched accurately in respect of all its tissues with authentic reference material. Descriptions and drawings are rarely full enough for one to be absolutely certain of a match. We have examined roots suspected to be the food supply of underground larvae of various insects. Only by having samples of roots from all the plants growing in the region where the larvae were found was it possible to make identifications of the chewed fragments. The division into monocotyledon or dicotyledon roots is relatively simple – it is only after that that the real problems begin!

Fortunately, secondarily thickened roots of dicotyledons (all monocotyledon roots are primary) are much simpler to identify. The economic significance of this is discussed on pp. 95–96.

Where to find particular characters

Leaf

Anisocytic stomata Cruciferae, Plumbaginaceae.
Anomocytic stomata Berberidaceae, Capparaceae, Liliaceae, Polygonaceae, Runanculaceae.
Branched or dendritic hairs Melastomataceae, Solanaceae.
Calcified hairs Boraginaceae.
Capitate glands Convolvulaceae, Labiatae, Sapindaceae.
Diacytic stomata Acanthaceae, Caryophyllaceae.

Hairs of varied types Compositae, Labiatae, Malvaceae, Solanaceae.

Hydathodes Campanulaceae, Piperaceae, Primulaceae.

Hypodermis Lauraceae, Piperaceae.

Latex-containing cells Apocynacaea, Convolvulaceae, Papaveraceae.

Mucilagenous epidermis Malvaceae, Salicaceae.

Papillose lower epidermis Berberidaceae, Lauraceae, Papilionaceae, Rhamnaceae.

Papillose upper epidermis Begoniaceae, Melastomataceae.

Paracytic stomata Gramineae, Juncaceae, Magnoliaceae, Rubiaceae.

Peltate hairs Bombacaceae, Oleaceae.

Salt glands Frankeniaceae.

Scales Bromeliaceae.

Sclereids Oleaceae, Theaceae.

Silicified hairs Gramineae.

Stinging hairs Euphorbiaceae, Loasaceae, Urticaceae.

Tufted hairs Fagaceae, Hamamelidaceae.

T-shaped hairs Malpighiaceae, Sapotaceae.

Stem

Cluster crystals Bombacaceae, Cactaceae, Chenopodiaceae, Malvaceae, Rutaceae, Tiliaceae, Urticaceae.

Cortical bundles Araliaceae, Cactaceae, Cucurbitaceae, Melastomataceae, Proteaceae.

Cystoliths Cannabidaceae, Moraceae, Urticaceae.

Deep seated cork Bignoniaceae, Casuarinaceae, Hypericaceae, Rosaceae, Theaceae.

Intraxylary phloem Apocynaceae, Convolvulaceae, Cucurbitaceae, Lythraceae.

Medullary bundles Begoniaceae, Compositae, Papilionaceae, Piperaceae, Saxifragaceae, Umbelliferae.

Primary medullary rays, narrow Asclepiadaceae, Cruciferae, Ericaceae, Meliaceae, Oliniaceae, Rubiaceae, Sapotaceae.

Primary medullary rays, wide Begoniaceae, Compositae, Cucurbitaceae, Ficoideae, Nyctaginaceae, Papilionaceae.

Raphides Balsaminaceae, Liliaceae, Rubiaceae.

Secondary thickening from multiple cambia Amaranthaceae, Chenopodiaceae, Menispermaceae, Nyctaginaceae.

Solitary crystals Flacourtiaceae, Mimosaceae, Papilionaceae, Rutaceae, Tamaricaceae.

Superficial cork Compositae, Corylaceae, Fagaceae, Labiatae, Meliaceae, Protiaceae, Umbelliferae.

Note: not all members of the families named above will show the features mentioned, but the features regularly occur where indicated. There are, of course, many other examples of families where features in these lists also occur.

Some leaf and stem characters to be found in common plants from various parts of the world

Since students in various parts of the world may find it difficult to obtain material of some of the text examples, an additional list is given here of plants which are either common in their country of origin, or have been cultivated in various parts of the world. If the particular species mentioned is not available, it will be worth looking at other species of the same genus, because they may show the same characters.

These short notes mention only a few of the interesting characters that may be seen in each species.

Abrus precatorius Leguminosae Stem: superficial cork, fibres in cortex and phloem, rhombic crystals, phloem wide, with inflated rays, broad intrusions of phloem into xylem, wide vessels solitary and in long radial chains, narrow vessels and tracheids in clusters, rays heterocellular and 1–6 cells wide, parenchyma of xylem aliform and in tangential bands, pith sclerified.

Aerva lanata Amaranthaceae Leaf: various hair types, stomata anomocytic and present on both surfaces, cluster crystals. Stem: phloem fibres, large crystalliferous cells in cortex, anomalous vascular tissue with succession of collateral bundles from cambial tissue, vessel elements with simple perforation plates and alternate intervascular pitting.

Aesculus hippocastanum Hippocastanaceae Leaf: hairs unicellular and short uniseriate with warty walls, stomata anomocytic, petiole vasculature composed of cylinder enclosing amphivasal bundles, tanniniferous cells, cluster crystals.

Ageratum conyzoides Compositae Stem: hairs, rounded cells of chlorenchyma, endodermoid sheath, fibre strands at phloem poles, vessels narrow in radial multiples with simple perforation plates.

Arbutus unedo Ericaceae Leaf: cuticle thick, stomata anomocytic, palisade chlorenchyma both adaxially and abaxially, sclerenchyma caps to vascular bundles, rhombic and other crystals, tannin in some abaxial epidermal cells.

Averrhoa carambola Oxalidaceae Stem: hairs thick-walled, unicellular; epidermis and hypodermis thick-walled, cortical fibres, vessel element perforation plates simple and oblique, tannin abundant, rhombic and cubic crystals abundant in cortex phloem and xylem.

Bidens pilosa Compositae Stem: polygonal in T.S., hairs, well developed endodermoid sheath, fibre caps to phloem poles, vessels narrow, in radial multiples, perforation plates simple.

Bougainvillea sp. Nyctaginaceae Stem: hairs short uniserrate, hypodermis, fibres at phloem-cortex boundary, vascular system anomalous, outer bundles embedded in thick-walled prosenchymatous tissue, inner bundles in parenchyma, rhaphide sacs in cortex and pith.

Briza maxima Gramineae Leaf: prickle hairs, silica bodies rectangular in epidermal cells with sinuous walls, stomata paracytic, sclerenchymatous girders opposite vascular bundles both abaxially and adaxially, bundle sheaths, inner sclerenchymatous, outer parenchymatous, chlorenchyma radiate.

Catalpa bignonioides Bignoniaceae Leaf: hairs peltate and uniseriate, stomata anomocytic and superficial, epidermal cells with sinuous walls.

Cistus salviifolius Cistaceae Leaf: hairs, non-glandular tufted and raised on mounds and glandular and capitate, stomata anomocytic, cluster crystals.

Coffea arabica Rubiaceae Stem: phloem fibres, vessels solitary with simple perforation plates, rays narrow, rhombic crystals and crystal sand.

Coldenia procumbens Boraginaceae Leaf: warty hairs with rosette of basal cells, stomata anomocytic, palisade adaxial, cluster crystals. Stem: collenchymatous outer cortex, vessels with simple perforation plates, rays narrow, pith of parenchymatous cells with conspicuous pits.

Cyperus papyrus Cyperaceae Stem: outline triangular in T.S., stomata paracytic, conical silica bodies in epidermal cells above hypodermal fibre strands, network of parenchyma with large air spaces, vascular bundles scattered and embedded in parenchyma.

Elaeis guineensis Palmae, Rachis petiole: vascular bundles in very thick sclerenchymatous bundle sheaths embedded in parenchymatous matrix. Lamina: hairs, expansion cells above and below midrib, hypodermis, spherical silica bodies.

Epacris impressa Epacridaceae Leaf: epidermal cells axially elongated with sinuous anticlinal walls, stomata anomocytic and superficial on abaxial surface only, vascular bundles with sclerenchyma caps at phloem pole.

Euphorbia hirta Euphorbiaceae Leaf: hairs, abaxial epidermal cells papillose, stomata anisocytic or anomocytic, laticifers, bundle sheaths with contents staining red in safranin, arm cells of spongy mesophyll clearly visible in paradermal preparations. Stem: vessels solitary or in radial multiples, perforation plates simple.

Fagus sylvatica Fagaceae Leaf: cuticle thin except over petiole, epidermal cells with sinuous anticlinal walls, hairs, stomata anomocytic and superficial on abaxial surface only, bundle sheaths with paired crystals, tannin abundant in cells of petiole. Stem: cork arising in outer cortex, phloem fibres, vessels diffuse porous and solitary or in pairs, perforation plates simple (scalariform in some narrow elements), rays uniseriate to multiseriate and heterocellular, xylem parenchyma scattered.

Gloriosa superba Liliaceae Leaf: epidermal cells over veins elongated with straight anticlinal walls, epidermal cells between veins with sinuous walls, stomata anomocytic abaxial, vascular bundle sheaths parenchymatous, spongy mesophyll composed of arm cells.

Hamamelis mollis Hamamelidaceae Leaf: hairs tufted consisting of 4–8 thick-walled radiating pointed cells sometimes raised on mounds, stomata superficial and anomocytic or tending paracytic, sclereids in mesophyll, large mucilage cells, cluster crystals, rhombic crystals, tannin cells.

Heteropogon contortus Gramineae Leaf: adaxial epidermal cells larger than abaxial, stomata paracytic, prickle hairs, silica bodies square to oblong to saddle-shaped, sclerenchyma in margins and as abaxial and adaxial girders to main vascular bundles, parenchyma bundle sheaths, chlorenchyma radiate. Stem: sclerified hypodermis, cylinder of fibres to inner side of cortex.

Hyphaene sp. Palmae Leaf: stomata appearing tetracytic, hypodermis, sclerenchyma bundle sheath extensions, fibre strands.

Lantana camara Verbenaceae Stem: hairs both glandular and non-glandular, phloem fibres, vessels with simple perforation plates, intervascular pitting alternate, rays narrow and heterocellular, xylem parenchyma abundant.

Mangifera indica Anacardiaceae Stem: cuticle thick, cortex with rhombic and prismatic and cluster crystals, tannin cells and cells with granular inclusions, phloem fibres, vessels angular and thin walled both solitary and in short radial multiples, perforation plates simple or a few scalariform, intervascular pitting coarse and alternate, rays 1–2 cells wide and heterocellular, axial secretory ducts lined with thin-walled epithelial cells in phloem and pith.

Nerium oleander Apocynaceae Leaf: cuticle thick, stomata and hairs in pits on abaxial surface, hypodermis, cluster and prismatic crystals, laticiferous canals near to veins.

Oxalis corniculata Oxalidaceae Stem: some epidermal cells containing tannin, complete cylinder of cortical fibres, vessel element perforation plates simple.

Pittosporum crassifolium Pittosporaceae Leaf: cuticle very thick, hairs, stomata paracytic with massive cuticular rim, sunken, hypodermis abaxially and adaxially, phloem poles to bundles disproportionately large, secretory canals of various diameters, cluster crystals.

Plantago media Plantaginaceae Leaf: hairs uniseriate and short, with bicellular head, stomata anomocytic and superficial, epidermal cells with sinuous anticlinal walls.

Polemonium coeruleum Polemoniaceae Stem: hairs, stomata slightly raised, outer cortex of rounded chlorenchyma cells, inner cortex collenchymatous, phloem with transverse sieve plates, vessels solitary or paired and diffuse and angular, perforation plates simple and oblique, intervascular pits fine and rounded, rays narrow.

Plumbago zeylanica Plumbaginaceae Leaf: glandular hairs, stomata, anisocytic, enlarged tracheids at vein ends.

Rubus sp. Rosaceae Stem: cork arising in middle cortex, suberized alternating with unsuberized layers, phloem fibres, primary rays broad, secondary rays 1–2 cells wide and heterocellular, vessels wide, in radial or tangential multiples, perforation plates simple, intervascular pitting alternate, pith composed of large and small parenchyma cells, cluster and rhombic crystals present in cortex and pith.

Salvadora persica Salvadoraceae Stem: epidermal cells of uneven heights and some raised into mounds, phloem fibres, xylem with included phloem, vessel element perforation plates simple, intervascular pitting alternate.

Sphenoclea zeylanica Sphenocleaceae Leaf: epidermal cells papillate, adaxial palisade chlorenchyma, parenchymatous bundle sheaths, cluster crystals. Stem: cortex with air-spaces, phloem fibres, vessel elements with simple perforation plates, intervascular pitting alternate.

Tamarix gallica Tamaricaceae Stem: cork superficial with large cells, phloem fibres, vessels solitary or in small radial multiples, perforation plates simple, rays 1–3 seriate and conspicuous composed of wide cells, crystal sand and irregular crystals abundant.

Tecoma capensis Bignoniaceae Stem: cuticle thick, hairs unicellular, cork superficial and to outer side of chlorenchyma, cortical fibre caps and strands of phloem fibres alternating with soft tissue, innermost fibres forming interrupted ring, phloem appearing storied, xylem with narrow vessels in solitary or in short radial multiples, vessel walls thick, perforation plates simple and oblique, intervascular pitting coarse and alternate.

Theobroma cacao Sterculiaceae Stem: hairs unicellular and thick-walled, mucilage cavities (canals) in cortex, phloem fibres, vessel elements with simple perforation plates, intervascular pits coarse and alternate.

Further reading

Carlquist, S., 1961. *Comparative Plant Anatomy*, Holt, Rinehart and Winston, New York.

Metcalfe, C. R. & Chalk, L., 1950. *Anatomy of the Dicotyledons* vols I & II, Clarendon Press, Oxford.

Metcalfe, C. R., editor, *Anatomy of the Monocotyledons:*
I *Gramineae*, 1960. Metcalfe, C. R.
II *Palmae*, 1961. Tomlinson, P. B.
III *Commelinales-Zingiberales*, 1968. Tomlinson, P. B.
IV *Juncales*, 1968. Cutler, D. F.
V *Cyperaceae*, 1971. Metcalfe, C. R.
VI *Dioscoreaceae* 1971. Ayensu, E. S.

5 Meristems

Growth takes place in two stages in plants: first there is the division of cells of an undifferentiated type, adding to the number of cells; then there is the enlargement of some of the cells produced by these divisions.

Dividing cells of the undifferentiated type are not present throughout the plant, but are concentrated in particular places. In addition to these, certain cells in most organs remain relatively undifferentiated and may begin to divide if the appropriate conditions arise and after they have undergone a process known as dedifferentiation. Such cells give rise to adventitious roots and buds, or to the callus tissue which forms during wound healing. They are of great importance to the horticulturalist. The ability of such cells to divide is a basic requirement for the success of many forms of vegetative propagation and grafting.

Cells which divide actively to produce the primary plant body are associated together in meristems. These comprise the *apical meristems* at the tips of shoot and root and the tips of lateral shoots or roots. Some plants have active meristems just above and adjacent to most nodes; these are the *intercalary meristems*.

When secondary growth occurs, that is, growth in thickness, the *lateral meristems* are involved. The vascular cambium occurs in dicotyledons and gymnosperms and is the best known of the lateral meristems. Growth in thickness of stem and root causes the primary covering layer of the plant, the epidermis, to split. A secondary protective barrier between delicate tissues and the outer world is developed to replace the epidermis. It consists of layers of cork cells, derived from the specialized cork cambium or phellogen, also a lateral meristem.

In the dicotyledon leaf, cells continue to divide in various areas of the expanding lamina, some until the mature size has almost been attained, when they cease division and the products expand. Leaves in monocotyledons are different; most have a basal zone of meristematic tissue which continues its growth for long periods, until the mature leaf size has been reached.

Certain monocotyledons have secondary growth in stem thickness, although many of the larger ones do not – e.g. the Palms. The Liliaceae, with *Dracaena* and *Cordyline* and the Iridaceae, with *Aristea* serve as examples in which there is a special zone of meristematic cells in the outer part of the cortex. Here entire vascular bundles are formed, with new ground tissue between them.

It can be seen, then, that the growing plant is exceedingly complex, containing areas which are juvenile and have actively dividing cells close to other tissues which are fully formed and mature.

Apical meristems

There are detailed differences between the meristems at the apex of shoot and root of monocotyledons, dicotyledons, gymnosperms and lower plants. Three shoot apices are shown in Fig. 5.1.

Since the earliest observations, writers have attempted to classify the various layers of cells in apices which are visible in longitudinal sections. Classification of these layers has been made on the basis of the fate of cells derived from the distinct layers, or on the dominant planes of cell division apparent in the layers. For example, in the tunica and corpus theory, the tunica layers can be distinguished from the inner, corpus layers because their cell divisions normally occur in the anticlinal plane only. In the corpus, divisions are both anticlinal and periclinal. If a formal naming of layers must be made, the tunica–corpus system is more reliable than Hanstein's dermatogen–periblem–plerome system, in which the

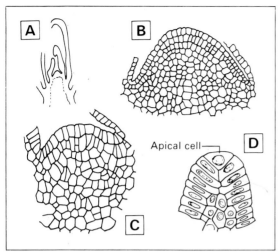

Fig. 5.1 Vegetative meristems. A, low power diagram L.S. of *Rhododendron* apex, × 15. B, detail of A, × 218. The second layer may be 'tunica' but has some periclinal divisions, as does that in C, *Syringa* × 218. D, *Equisetum* × 218, has an apical cell and not a group of meristematic cells.

layers are defined in relation to the tissue systems to which they are purported to give rise. However, since it has been shown by experiment that particular layers do not consistently give rise to the same tissue system in the same species, it is possibly better to use a topographic system and label layers L1, L2, L3 etc. and define various zones descriptively.

In the shoot apex, leaves usually arise from the tunica layers (L1, or L1 and L2 normally) and buds from tunica and some corpus layers. The tunica produces the epidermis and usually most if not all of the cortex. Sometimes cell divisions occur very early in the epidermis during the development of leaves, leading to the production of a multiple epidermis. This can be seen in Fig. 5.12, part of the apex of *Codonanthe* sp. (Gesneriaceae). Part of the mature multiple epidermis of this plant is shown in Fig. 7.7. The corpus produces the vascular system of the stem and the central ground tissue. Occasionally the cells below the apical meristem proper may appear to be relatively inactive in terms of division; this region is termed the quiescent zone, but its inactive state is not acknowledged by all, and experiments using radioactive tracers indicate that there is some cell division in these regions. A regular rib-like arrangement of cells can also be detected in some apices below the tunica and corpus. The cells of the meristem have dense cytoplasm, lacking large vacuoles. Below the areas of active cell division, the cells begin to elongate and vacuolate.

Several types of organization of cell zones or layers are recognized by various authors as being present in the gymnosperms. The majority of angiosperms appear to conform to one type, and only a small minority to a second. There must be a very large number of species that have not been investigated for the type of apical arrangement, and it would seem probable that additional types could exist. It can be seen that the interpretation of apical organization is a complex and controversial subject. The interested reader is referred to the 'Further reading' at the end of this chapter.

As leaf buttresses arise in sequence at the apex, in the phyllotaxy characteristic of the particular species, the procambial strands become apparent, and from them are derived the first formed phloem and then the xylem of the primary bundles. Figure 5.11 shows leaf buttresses in *Elodea* as seen in the S.E.M. photograph taken by Dennis Stevenson. Many experiments have been conducted to try to find out the mechanisms which regulate the orderly development of these dynamic, growing apices. Control of spacing of leaf buttresses is not fully understood. Numerous experiments involving the use of mechanical devices to try and isolate one part of the apex from the rest have been carried out, and despite these painstaking experiments with growth hormones there is still a great deal to learn. It is very difficult to conduct experiments in which only one variable at a time is studied. Also, apices develop in the very enclosed, protected environment of the leaf bases which has to be substantially disturbed so that observations can be made.

The root apex is similar in many respects to the stem apex and may also have a quiescent zone, but it has one conspicuous, major difference. It has a root cap, or calyptra, frequently produced by a meristematic zone called the calyptrogen (Fig. 5.2 A). The cap acts as a buffer between the soft apical meristem and harsh soil particles. It wears away as growth progresses, but it is constantly renewed. It is believed to be the source of growth regulating substances which are involved in the positive geotropic response of most roots. Root caps can be seen easily on aerial roots of *Pandanus* and many epiphytic orchids. Besides the calyptrogen, the cell layers responsible for production of root epidermis and cortex, and the primary vascular system can be readily defined in thin, longitudinal sections suitably stained. In some roots no distinct calcyptogen is produced. In *Allium* (Fig. 5.2 B, C) a column of cells develops.

Whereas the shoot apex soon produces leaves and buds exogenously, the root organization is quite different. Lateral roots arise endogenously, from the

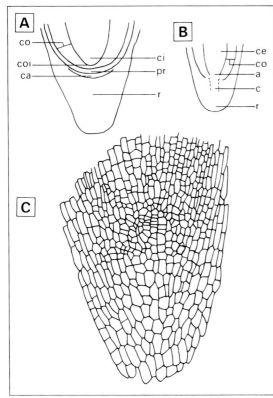

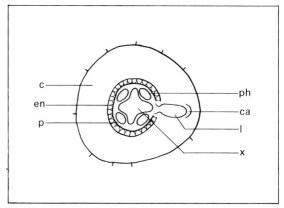

Fig. 5.3 Endogenous development of a lateral root, in T.S. c, cortex; ca, small cavity ahead of developing lateral root formed by lysis of cortical cells; en, endodermis; l, lateral root; p, pericycle; ph, phloem; x, xylem.

Fig. 5.2 A, generalized monocotyledon root, diagram to show location of various zones. Root apex, L.S., in *Allium* sp.; B, low power diagram to show location of various cell zones in C (× 109). a, apical meristem; c, column; ca, calyptrogen; ce, central cylinder; ci, central cylinder initials; co, cortex; coi, cortex initials; pr, protoderm initials; r, root cap.

pericycle cells, some distance from the apex (Fig. 5.3). This deep seated origin necessitates that the lateral roots grow forcibly through the endodermis and cortex to reach the exterior. The distance to be spanned between the lateral root vascular system and that of the root from which it arises is short. The vascular system from a bud apex has to develop towards the main stem system and eventually becomes joined to it.

Numerous papers have been published on the apical organization of shoots and roots in many plant families. Some are comparative and aim at conclusions of taxonomic significance, but others, and probably the more useful, are concerned with the development of the particular plants under study. As mentioned before, proper developmental studies demand a high degree of competence, and are vital to an understanding of mature plant forms.

Applications

Apical meristems

The main practical use to which meristems are put – particularly shoot meristems – is meristem culture. This is a method of vegetative multiplication which requires careful excision of the apex and its culture or growth on a nutritive medium. All stages of this technique must be aseptic so that no pathogens are introduced. If the process is successful the apex will first form a mass of callus-like tissue, similar to the orchid protocorm. Then small embryonic shoots and roots appear. If the tissue mass is subdivided, then a number of small plantlets can be produced. It is important to have a correctly formulated growth medium often peculiar to the plant under culture, or the tissue mass may produce only shoots or roots!

There are several circumstances where it is desirable to reproduce plants by meristem culture. For example, the required plant may be infertile, as in the case of a triploid, or it could be an F1 hybrid which would breed untrue. It is also a useful method for the rapid increase of nursery stock for commercial purposes. Other vegetative methods of propagation might take several additional years before a similar number of plants could be produced. Virus diseases rarely infect apices, and meristem culture can be used to produce virus free stock from otherwise infected plants, for example in the raspberry and the potato.

As a method of propagation, meristem culture would appear to have a bright future. It probably has more potential than the longer established callus culture method, whereby small portions of excised tissue (usually parenchyma) from various parts of a plant are cultured in, or on, a nutritive medium. It may take a long time to induce embryo plants to differentiate from such a callus.

When large enough to handle, the embryonic plants are detached and grown on a sterile medium to a size at which they can be potted on, in a normal potting compost.

Intercalary meristems

Intercalary meristems are also used in horticulture for propagation. In the plant one of their functions is to cause a stem which has fallen over to grow back up-right again, for example in *Triticum* or Carnation, Carnations will serve as a practical example of where an intercalary meristem is capable of producing adventitious roots. In Fig. 5.4 a carnation stem is shown cut off the plant just below a node. It is split longitudinally through the node, into the intercalary zone. In horticultural practice the split is often held open with a piece of stick. Adventitious roots develop from the split sides.

A large number of plants quite readily form adventitious roots from the nodes, whether split off the plant or not. Considerable use is made of this property in horticulture for propagation.

Lateral meristems

Lateral meristems are also used in techniques designed to propagate plants by cuttings, and also in grafting.

The cork cambium is so specialized as to be of little value in plant propagation. It frequently plays a part in wound healing, and is, of course, employed commercially in the production of 'cork' from *Quercus suber*, the cork oak, in which the cork layers are harvested about every ten years. A new cork cambium or phellogen forms after the cork is carefully removed. Figure 5.5 shows a cork cambium in *Ribes nigrum*.

Of the lateral meristems, it is the cambium between phloem and xylem which is most often employed by horticulturalists. Its normal function in the healthy woody or herbaceous dicotyledon is to produce new cells of phloem and xylem (see Chapter 6). It consists of a layer of thin-walled cells situated initially in the fascicular region but in most dicotyledonous plants soon extends between the vascular bundles to form a complete cylinder. By periclinal divisions it forms new

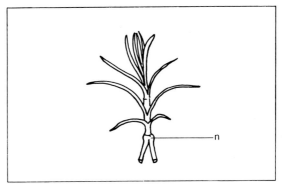

Fig. 5.4 *Dianthus* (carnation) cutting. Adventitious roots will develop from the split sides. n, node.

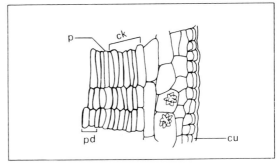

Fig. 5.5 *Ribes nigrum*, T.S. of sector of outer part of stem to show deep seated cork cambium, ×218. ck, cork; cu, cuticle; p, phellogen; pd, phelloderm. Note cluster crystals in cortical cells.

phloem cells to the outer side and new xylem cells to the inner side, whilst remaining intact itself. Normally, many more of the products of such divisions are xylem than phloem cells. Cambial initials are of two kinds, the fusiform initials which form vessels, tracheids, fibres and parenchyma of the axial parenchyma, and the much smaller ray initials which produce ray cells of the radial system (Fig. 5.6).

It is often very difficult to locate the actual cambial layer because it is composed of thin-walled cells, and the cells which are divided from it have thin walls for a while. As new layers are formed, they cause displacement of older phloem layers. Formation of new layers of xylem effects the displacement of the cambium itself away from the centre. Some of the cambial initials divide anticlinally to allow for the necessary increase in circumference. Also, occasional fusiform initials divide to form new radial initials and so maintain a more or less constant proportion of rays per unit volume of new xylem and phloem. This proportionate relationship is often fairly constant for a particular species.

If a cambium is wounded, it will normally regenerate and by influencing the developmental path-

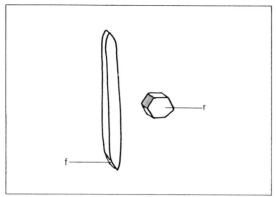

Fig. 5.6 Diagram of fusiform cambial initials (f) and ray initials (r).

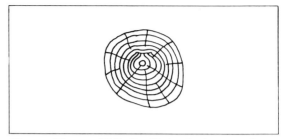

Fig. 5.7 Diagram of T.S. of conifer log showing resumed continuity of growth rings after lateral branch has been cut off.

ways of callus cells adjacent to it, will assist in the healing process, so that cambial continuity is regained, and new cylinders of phloem and xylem established.

The forestry practice of removing lower branches on conifers at an early stage enables the wound to heal over (Fig. 5.7) so that entire rings of sound new wood may become established. If 'snags' or broken ends of branches are left, it is some time before they are grown over, and bad knots result.

The inherent ability of wounds to heal is widely used in grafting techniques. In order to establish a graft, the parts of the two plants to be united are 'wounded'. This is done by cutting the root stock and scion or bud. The two are brought together in such a way that the cambia of stock and scion are as closely aligned as possible. When new growth is formed by callus cells the two cambia can then quickly establish continuity by specialized differentiation of some of the callus cells and a firm, even bond is produced. No cell fusion takes place, but eventually the xylem products of the two (joined) cambia firmly bond stock and scion together (Fig. 5.8). It is essential that the stock and scion should not be able to move relative to one another during the early stages, and in these stages grafting tapes are used which both give a secure bond and permit diffusion of oxygen essential to cell growth. The tape must either perish by itself in due course, or be easily removed by one cut. After-care is very expensive for the nurseryman. The simpler the method and the less handling involved, the better. Air gaps between stock and scion must be avoided; they can harbour pathogens, or permit the entry of water.

Bud grafting works in much the same way, and the chip bud graft (Fig. 5.9) is becoming increasingly popular and replacing the older T-cut method. The bud cambium can be aligned more accurately by this newer method. A chip bud is removed and inserted behind the small lower lip of bark of the stock at the depth of the cambium and the bud secured by grafting tape.

The advantages of grafting are manifold. For instance the roots of some desirable species may be very weak, and vigorous roots can be grafted in their place as in *Juniperus virginiana* where *J. glauca* stock is employed. Water melon with wilt-prone roots can be grafted onto a gourd root stock which is *Verticillium* wilt resistant. The sizes of fruit trees, particularly apples and pears, can be regulated by careful selection of root stock vigour. Trees of relatively fixed mature size can be produced, and earlier fruiting induced. The Malling-Merton system provides trees with numbered root stocks guaranteeing a mature tree with specific characteristics. Uniformity of size is essential for good husbandry. The dwarfing stocks have xylem vessel elements which are much narrower in diameter than those of the stocks producing large trees.

Bridge grafts can be used to repair trees which have been ring barked (Fig. 5.10). It is important to use twigs from the same species, since compatibility between stock and scion is essential. In fact, the inter-relationship between plants can be tested to a limited extent by their inter-grafting ability. Species from the same genus will frequently unite, e.g. *Prunus* species. *Solanum* species can also be grafted together. Graft hybrids between genera are much less common, e.g. *Laburnum/Cytisus*. Grafts between plants from different families probably do not occur.

Bud grafts are used to propagate material rapidly, e.g. to bring a new rose onto the market quickly. Roses, particularly hybrid teas and floribundas are often poor performers on their own roots, and of course will not come true to type from seed. In such instances grafting onto a healthy vigorous root stock performs the dual function of providing a vigorous root and helps in rapid propagation.

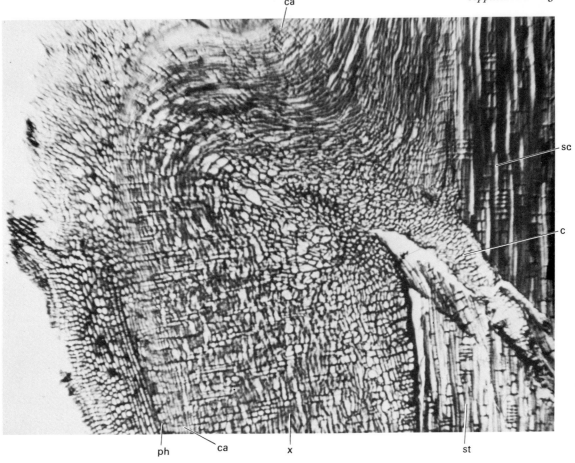

ca

sc

c

ph ca x st

Fig. 5.8 *Camellia* graft, L.S. of one side showing new xylem and phloem formed from cambium developed in the callus tissue. The new cambium connects that of stock and scion, uniting the two; there is no cell fusion. Note the small remaining wedge of callus cells (c). ca, cambium; ph, phloem; sc, scion; st, stock; x, xylem.

If the vigour of the scion greatly exceeds that of the stock, ugly overgrowth can occur, and where there is no desire to regulate scion vigour, a stock of suitable vigour should be selected.

The callus cells produced by wounding two (or more) plants can sometimes be grown in culture, and groups of cells centrifuged together. The resulting complex of cells can be grown on, producing cytohybrid plants of the most complex type of graft imaginable – that is, except for the fusion of protoplasts of two different organisms which represents the extreme form of grafting!

Monocotyledons are virtually impossible to graft, although there are a few reports of such grafts in the literature. Most have no secondary growth in thickness. The vascular bundles are 'closed' and produce no cambium. Some appear to have cambial division, but this may merely be the late, rather regular divisions of cell layers which take place in the central region of the bundle when it is approaching maturity. However, not enough is known about this phenomenon, and it remains an open question whether grafts can become established in monocotyledons.

The monocotyledonous bundle is often firmly enclosed by a sclerenchyma or parenchyma sheath, or both. So the monocotyledonous bundle lacks the meristematic cells needed to effect fusion, and the accurate positioning of bundles in a graft would, furthermore, be virtually impossible.

As mentioned earlier, secondary growth in thickness does occur in some monocotyledons, but it is brought about by special tissues at the periphery of the stem. These are in effect a lateral meristem and produce both new, entire vascular bundles by cell division and the new ground tissue between the bundles. *Cordyline* has the type of secondary thickening common to a number of monocotyledons. Again it is easy to see that grafting would fail because

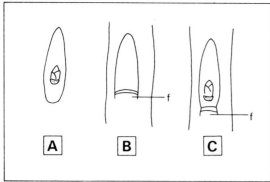

Fig. 5.9 Chip bud graft. A, chip with bud; B, stock prepared; C, chip inserted behind small flap of bark (f), ready for taping.

Fig. 5.10 Twigs grafted across a damaged area of bark on a tree trunk.

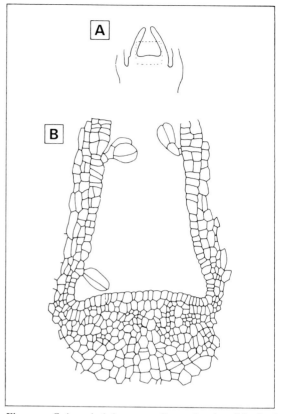

Fig. 5.12 *Codonanthe* A, low-power diagram location of details of apex of shoot shown in B. Notice the very early division of cells in the adaxial epidermis of the leaves, leading to the formation of a multiple epiermis. B, × 248.

Fig. 5.11 Shoot apex of *Elodea* showing developing leaf buttresses, S.E.M. × 100, live material.

it is not possible to align enough bundles and little or no vascular continuity can be achieved.

Adventitious buds

Some plants have the ability to produce adventitious buds from various organs when the plant or part of the plant is placed under some unusual physiological stress. The stress may be caused by an injury or even by the separation of one organ from the rest of the plant. The development of buds in this way is usually thought to be related to the loss of a constraint, for example, the loss of some inhibitory hormone or similar chemical substance. When the apical dominance of a shoot system is removed, new adventitious buds (not related to leaf axis) may develop. This enables us to lop certain species of mature tree and get new growth. For example *Salix* and *Platanus* will grow new branches from adventitious buds. The practice of pollarding and harvesting the pole-like young branches would not be possible if this type of recovery did not take place. Some crops such as *Quillaja* and *Cinchona* bark are obtained from coppiced trees; extracts from these are used in the preparation of medicines.

Further Reading

Cutter, E. G., 1965. 'Recent experimental studies of the shoot apex and shoot morphogenesis', *Botanical Reviews*, **31**, 7–113.

Wardlaw, C. W., 1968. *Morphogenesis in Plants. A contemporary study*, Methuen, London.

Williams, R. F., 1975. *The Shoot Apex and Leaf Growth*, Cambridge University Press.

See also list of general advanced texts, p. 14.

6 Xylem and phloem: the secondary systems

Xylem

Secondary xylem, or wood, is put to an extremely wide range of uses. The extensive range of species from the gymnosperms and angiosperms which are used as sources of wood is reflected in the diverse properties of the various kinds of wood.

There is archaeological evidence that our early ancestors were well aware of the best woods for burning for warmth or metal smelting, those most durable and strong for making boats or buildings and those most suited as shafts for tools or weapons. They even selected carefully for their musical instruments and decorative carvings. In our advanced stage of technology we make use of the different characteristics of strength, workability, durability, density and pulping potential in our selection of woods for a vast range of primary and secondary products.

Obviously, this wide range of properties must be accounted for by variations in the histology and fine structure of woods. In fact there are many characters in which wood can vary but it is not always clear what effects they have on the properties of the wood. The possible variation is so great that the set of characters shown by wood from a particular species can provide clues to the identity of the species. Sometimes the set of characters may indicate only the family or genus but occasionally they are confined to a species. In other words, one would expect individuals of the same species to share very similar wood characters, but another closely related species might be so similar that it cannot be distinguished by wood features alone.

In this chapter we shall explore the sorts of differences which occur in wood, and look at the ways in which these help in identification and in establishing the relationships between species, and how they affect the properties of the timber.

Evolution in secondary xylem

It is generally accepted that secondary xylem has undergone a long evolutionary history. The main trends can be seen because the various stages are often related to other 'marker' characters in flowers, fruits etc. of the plants concerned. There are instances where habitat has seemingly reversed some of these trends in various species, but overall, their 'direction' can be fairly safely defined.

Taken in its simplest form, the evidence to hand indicates that the tracheid, a dual-purpose cell combining properties of both mechanical support and water conduction in evolving groups of plants, gave rise to fibres with simple mechanical function and to perforate cells (Fig. 6.1), the vessel elements concerned with the conduction of water and dissolved salts. This division of labour is seen as a specialization, or advance.

The primitive vessel element shows much similarity to the tracheid; it is axially elongated, with oblique end walls in which are grouped perforations making up the scalariform, reticulate or otherwise compound perforation plates. The lateral walls bear bordered pits, often in an opposite arrangement. The advanced vessel element is seen as a broad short cell with large, simple transversely arranged perforation plates at either end and alternating bordered pits on the lateral walls. Between these extremes is a variety of forms (Fig. 6.2).

In the monocotyledons, the vessel element probably evolved first in roots, then in stems and finally in leaves. Evidence for this is found in many plants. There is no record of a species being found with vessels in the leaves only, and not in the roots.

There are some flowering plants which are considered to be primitive because of floral characters, and which are vessel-less (for example *Drimys*, Magnoliales, among the dicotyledons).

Confidence in the evolutionary sequence is such that characters of the vessel elements have often been

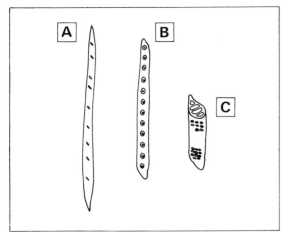

Fig. 6.1 A, fibre, B, tracheid and, C, vessel, contrasted;
intermediate cell types exist between each.

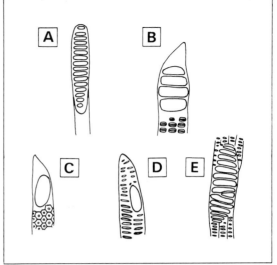

Fig. 6.2 A range of vessel element perforation plates and wall
pitting, all × 218. A, *Camellia sinensis*, scalariform. B,
Liriodendron tulipifera, scalariform; pits opposite. *Sambucus nigra*,
simple plate, pits alternate. D, *Euphorbia splendens*, simple plate,
pits opposite. E, *Scirpodendron chaeri*, sclariform plate, pits
opposite (from primary xylem).

used as an indicator of the relative phylogenetic
advancement of plants. Measurements of vessel element length and width must be made on a statistically
sound basis for such comparisons. It has been found
that a ratio of vessel element length to tangential
width produces a useful figure for application in
advancement indices. There are many pitfalls in this
method. Great care must be taken in ensuring that
comparisons are made between plants growing in
fairly similar conditions, since habitat can influence
the vessel diameter. There can also be a degree of
natural variability which must be accounted for in
sample size. Comparisons between the trends in
families or genera are naturally more reliable than
those between species. Overall trends in orders are
again of more significance. So, even in the detailed
measurements of fibres, tracheids and vessel elements we see features which can be applied in evolutionary terms, or used when we are attempting to
establish the possible origin of a group of plants. For
example, it would be unlikely that plants with vessels
in root, leaf and stem would be ancestral to those with
vessels in the root only. These data are of considerable
interest to phylogenists.

The construction of secondary xylem

Primary xylem has been described in Chapter 4. It
consists of the axial cell system only, that is, xylem
cells which are elongated parallel with the long axis of
the organ or vascular trace in which they occur.
Secondary xylem, one of the products of the vascular
cambium, is more complex. As we have seen in
Chapter 5, the cambium is composed of two sorts of
cell, the axially elongated fusiform initials which give
rise to the axial system of cells and the short, more or

less isodiametric initials giving rise to the radial
system or rays. Figure 6.3 shows the axial and radial
systems in *Alnus glutinosa* wood.

Because both axial and radial systems are present,
the study of secondary xylem can only be carried out
properly by examination of three specific planes of
section from a block of wood. These are the transverse section (T.S.) the radial longitudinal section
(R.L.S.) and the tangential longitudinal section
(T.L.S.). These expose details of both systems of
cells. Figures 6.4 and 6.5 show these planes of section.

Gymnosperm wood (conifer)

In conifer woods (and gymnosperms generally) the
axial water conducting system is composed largely of
tracheids. There are no vessel elements. These
tracheids are rather like elongated boxes, with a
rectangular cross section and tapering upper and
lower ends. They communicate with one another
mainly through bordered pits in the lateral (radial)
walls. The size, number of rows and details of pit
structure are often characteristic for given species or
genera. Figure 6.5 shows bordered pits in *Pinus*, and
the glossary (see Pit, bordered) gives a reconstruction
of the bordered pit pair between adjacent tracheids.
Tracheids formed during the flush of spring growth
are usually wider radially than those formed later in
the growing season. It is normally easy to see the
extent of thickness of a growth ring for this reason

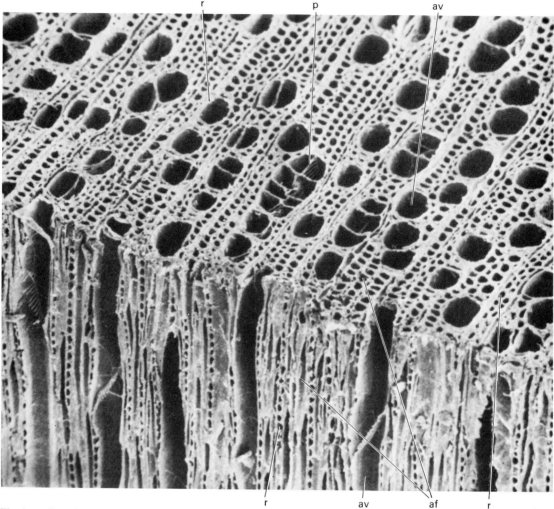

Fig. 6.3 *Alnus glutinosa*, SEM photograph of secondary xylem showing transverse and tangential longitudinal faces. af, fibres in axial system of cells; av, vessel of axial system; p, perforation plate (scalariform); r, uniseriate ray of radial system, × 100.

(Fig. 6.4). Sometimes a band of spiral thickening occurs inside the secondary wall of the tracheid. This is characteristic of *Pseudotsuga* and *Taxus* and some *Picea* spp., *Cephalotaxus* and *Torreya* in the mature trunk wood. However, many conifers have such a tertiary spiral thickening on tracheid walls in twig wood, so if narrow diameter specimens are to be identified this must be borne in mind. In badly decomposed wood, spiral splits may appear in tracheid walls, following the alignment of cellulose microfibrils in one of the wall layers; these can be confused with true spiral thickenings.

Fibres are not normally found in conifer woods, and axial parenchyma is rare. Members of the Pinaceae (except *Pseudolarix*) and *Sequoia* spp. of Taxodiaceae, have vertical resin ducts. These ducts are lined by a secretory epithelial layer of cells which remains thin-walled in *Pinus* species (Fig. 6.4) but become lignified in the other genera. The presence of resin ducts is, then, of taxonomic significance in the Coniferae. Some genera, like *Cedrus* may have traumatic ducts which must not be confused with those of Pinaceae.

The radial system in gymnosperms consists of parenchymatous cells which are, on the whole, procumbent. In some species there may be radial tracheids as well, and some Pinaceae have radial resin ducts. The part of the ray cell wall bordering onto the tracheid is usually pitted; the pits in this cross field area are commonly characteristic for the genus or for a group of genera and can be used diagnostically. Figure 6.6 shows some such pit types. The wall pittings and thickenings of ray tracheids may also be of a characteristic form, for example, the 'dentate'

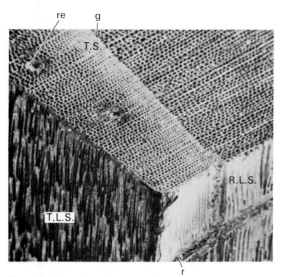

Fig. 6.4 *Pinus sylvestris*, corner of cube of secondary xylem as seen in SEM T.S., transverse surface; T.L.S., tangential longitudinal surface; R.L.S., radial longitudinal surface. g, growth ring, on one side the wider cells of spring wood, on the other the narrower cells of late season growth; r, ray; re, resin canal of axial system, × 30.

ray tracheids of *Pinus* species. Rays are normally one, sometimes two cells wide.

Detailed variations between genera are too numerous for mention in this text, but useful references are given at the end of the chapter for further reading.

Angiosperm wood

In dicotyledonous hardwoods we have noted that there are more types of cell to be considered, and their appearance and distribution in the wood give rise to a great range of wood types. Although details of wall pitting still remain important in the identification of specimens, there are many more characters which are readily observable in dicotyledonous woods than there are in conifer woods. Families or groups of families often exhibit quite characteristic features for study and comparison.

In the axial system of dicotyledons, tracheids are usually relatively sparse, most of the cells being vessel elements and fibres or fibre tracheids with varying amounts of axial parenchyma. These are all derivatives of the fusiform cambial initials. Fibres can be longer than the initial from which they were cut. Many fibres are capable of elongation by apical intrusive growth. Axial parenchyma cells are usually shorter than the initials which gave rise to them because the derivative cells often divide twice, to

form a chain of four cells. Xylem parenchyma cells normally have somewhat thickened, lignified walls. Vessel elements are extremely variable in their mature form, but particular forms are fairly consistently present in a given species. Details of wall pitting and perforation plates are illustrated in Fig. 6.2. Each of these features may be used diagnostically. Tertiary spirals may occur; an example is shown in *Tilia* in Fig. 6.7.

Rays are far more complex and show a wider range of variation in dicotyledons than in gymnosperms. They do not contain tracheids, but the parenchyma cells of which they are composed may exhibit a range of shapes and sizes. In some woods, e.g. *Castanea*, *Lithocarpus*, the rays are all uniseriate. In others, e.g. *Ulmus*, *Fagus*, they may be from one to several cells wide. In other species, e.g. *Quercus*, they are of two distinct sizes, some uniseriate and others wide and multiseriate, with no intermediates (Fig. 6.8).

In radial view, ray cells appear as courses of bricks in a wall. In some species all the cells are of similar size and proportions (homocellular); in others distinctly recognizable differences in cell size may occur (heterocellular). Cells of any particular size or shape are usually arranged in regular 'courses' or may be in particular positions, e.g. at the top and bottom of a ray. A range of different ray types is shown in Fig. 6.9, and includes procumbent and upright types; see also Fig. 6.16B.

With the possible combinations of cells and rays it is easy to see how different particular woods may be from one another. Wall thickness also has a bearing on the density and hardness of a wood. For example balsa wood *Ochroma lagopus* is lighter than cork (Fig. 6.10 shows *Ochroma pyramidalis* wood in T.S.; note the thin-walled fibres); Lignum vitae, *Guaiacum officinale* is extremely dense and heavy (Fig. 6.11; note thick-walled fibres) and is used for making bowls and pulleys, among other things, and black ironwood, *Krugiodendron ferreum* is even more dense.

Some wood have their rays and fibres arranged in regular horizontal rows as viewed in T.L.S. This storied type of wood gives a particular 'figure' to planks, and is often of decorative value. Many of the Leguminosae are like this. Unfortunately, it is rarely possible to say which particular anatomical features of a wood make it suitable for specific mechanical uses. In ring porous woods, for example, it seems that the number of growth rings to the inch is often of relatively more importance than histological details.

Evenness of texture or straightness of grain are features which belong to certain woods. Lime, *Tilia* sp. and pear, *Pyrus* sp., for example, have properties which make them good for carving; the wood cuts

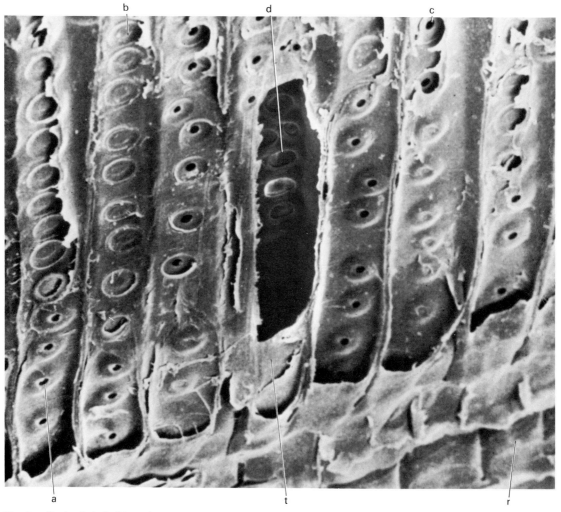

Fig. 6.5   Bordered pits in *Pinus sylvestris* wood tracheids as seen in R.L.S. Splitting the wood reveals the various wall layers. a, bordered pit in surface view; b, pit membrane with torus; c, inner side of pit cavity and pit aperture to next tracheid; d, pits between tracheid and ray cell walls (cross field pits). r, ray; t, tracheid, with part of wall torn away. SEM, × 300.

well in any direction. Ash, *Fraxinus* and hickory, *Carya* have a straight grain and are reselient, and are chosen for axe and tool handles. Long fibres or tracheids are one of the requirements for making certain types of paper. Soft woods (conifers) are usually preferred to hard woods for pulping for this reason.

Light woods with cells having moderately thickened walls are often more resilient and recover their shape better after denting than dense woods. The cricket bat willow, *Salix alba* var., *caerulea* is such a wood.

Many woods with good resistance to decay contain oils, gums or resins. Teak, *Tectona grandis* is a good example and was extensively used in boat building. *Bulnesia sarmienti* has gums and resins which produce an incense. *Cinnamomum camphora* is the source of natural camphor.

Spruce, *Picea* sp., has good resonating qualities and is widely used in the resonating chambers of stringed instruments. Oaks, *Quercus* spp., were extensively used since the time of iron age man in buildings and boats. Oak can be split using wedges, along lines of weakness formed by the broad rays, and planks or posts can be formed with simple tools.

A range of types of wood is illustrated in Figs. 6.12–6.16. These are chosen to demonstrate variations in vessel, fibre and parenchyma distribution and in ray type. Several excellent general books exist on wood anatomy, and there are many volumes on woods from particular parts of the world. Some of these are listed at the end of this chapter.

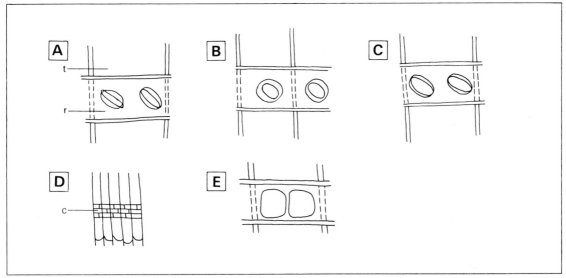

Fig. 6.6   Some types of cross field pits in conifers. A, piceoid, e.g. *Picea*, *Larix*. B, cupressoid, found in most Cupressaceae and *Taxus*. C, taxodioid, e.g. Taxodiaceae, *Abies*, *Cedrus*, some *Pinus*. D, diagram showing location of cross field pits (c) where ray and tracheid walls are adjacent. E, some *Pinus* spp. have large 'window' pits. r, ray; t, tracheid.

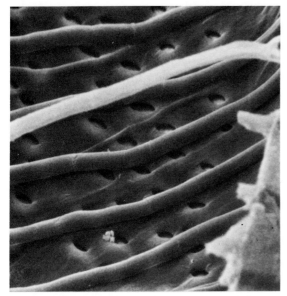

Fig. 6.7   *Tilia europaea*, L.S., tertiary spirals on vessel element wall, SEM, × 3,000.

Phloem

The primary phloem of monocotyledons and dicotyledons has been described in Chapter 4. As with secondary xylem, secondary phloem has both axial and radial arrangements of cells. The same initials in the cambium which divide to form xylem to their inner side also cut off phloem cells to their outer side. Sometimes growth rings can be seen.

Gymnosperm phloem
In gymnosperms, the axial phloem consists of sieve cells and parenchyma cells, some of which become albuminous cells (Fig. 4.29); some gymnosperms have fibres in the phloem as well. The homocellular rays are normally uniseriate. There is often very little wall thickening but sclerification can take place. The outermost phloem layers either become compacted, or are incorporated into the 'bark' or rhytidome.

Angiosperm phloem
The phloem cells of dicotyledons show evidence of evolutionary trends similar to those of the xylem. The sieve areas which are areas of dense pitting in lateral walls of sieve cells are a feature of the more primitive dicotyledons. Well organized sieve plates, simple and transverse, situated at either end of the sieve tube element are considered to be advanced. Oblique, compound sieve plates also occur (Fig. 6.17); sometimes these are found in advanced genera such as *Quercus* and *Betula*, and also in lianes – e.g. *Vitis* where physiological demands may call for large areas of sieve plate which are necessary for rapid, long-distance translocation of materials. Even in species where sieve plates are well developed, the lateral walls

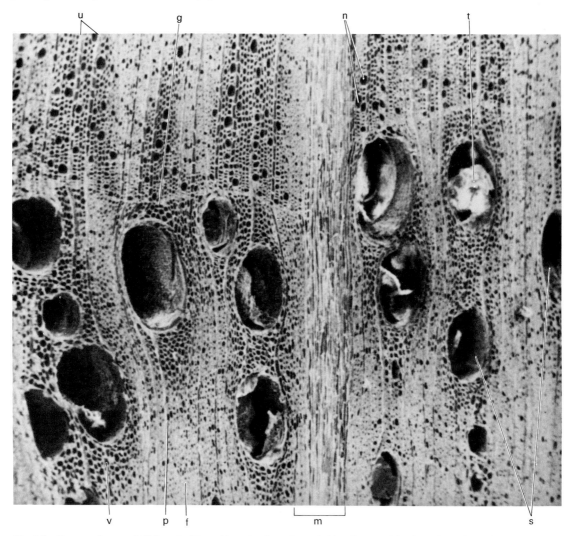

Fig. 6.8 *Quercus robur*, wood, T.S., ×60. Note wide spring formed vessels (s) and narrow later formed vessels (n). One multiseriate ray can be seen (m) and numerous uniseriate rays (u). Small parenchyma cells of the axial system occur in more or less tangential bands among the fibres (f). Tyloses (t) are present in some of the wider vessels (light appearance due to slight charging in the SEM). A growth ring (g) is shown, as is vasicentric parenchyma (v).

of the sieve tube elements often have distinct areas of pitting, called sieve areas. Companion cells, usually much narrower than the sieve tube element to which they are adjacent, are a feature of dicotyledon phloem. Their counterpart in the gymnosperms is thought to be the albuminous cell. Companion cells are nucleate, sieve tube elements are not. Killing a companion cell appears to prevent the adjacent sieve tube element from translocating, so the function of the companion cells seems to be in regulating the physiological activities of the sieve tube elements.

The axial system of secondary phloem often contains parenchyma, idioblastic cells, sclereids and fibres. In some species, sclereids and fibres are absent from the functioning phloem, but differentiate at a later stage. Fibres often alternate, in bands, with conducting cells, e.g. in *Tilia*, and various Malvaceae (Fig. 6.18). Primary phloem fibres of *Linum* (flax) are of economic importance.

The rays in phloem may be homocellular or heterocellular. In some species they remain of even width, but in others they may become wider towards their outer ends (e.g. *Tilia*). The rays may be uniseriate to multiseriate. As in secondary xylem, the secondary phloem may be storied; naturally, the storied arrangement originates from the storied

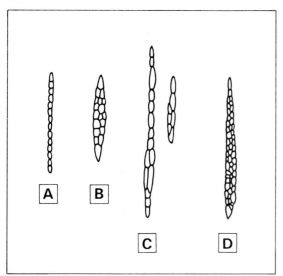

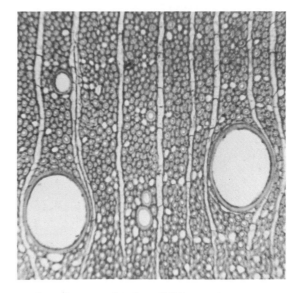

Fig. 6.9   Some ray types in T.L.S. all × 72. A, *Alnus glutinosa*, uniseriate, homocellular all cells of procumbent type. *Swietenia mahagoni*, multiseriate, heterocellular, with upright cells at margins and procumbent cells between. C, *Sambucus nigra*, biseriate, with tall uniseriate portions, heterocellular. D, *Musanga cecropoides*, multiseriate, heterocellular, procumbent and upright cells together in body of ray, upright cells at margins.

Fig. 6.11   *Guaiacum officinale*, wood T.S.; note the numerous thick-walled fibres and the scattered parenchyma, × 200.

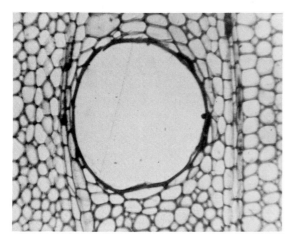

Fig. 6.10   *Ochroma pyramidalis*, wood T.S.; note the vessel, the thin-walled fibres and abundant parenchyma, × 200.

cambium in these plants. Laticifers and lysigenous cavities of various kinds may occur in the phloem.

The application of phloem anatomy in taxonomy has not been of such widespread interest as might be expected. The partially sclerified, often highly crystalliferous tissue is difficult to section well. This has put off many people! When phloem tissues form part of the rhytidome they have been more intensively studied than when they are distinct from the rhytidome, and a number of valuable contributions exist on 'bark' anatomy.

Most applied studies are concerned with the relationship between fine structure of phloem and its function. Relatively few plants are amenable to this sort of study, since accessibility of the functioning cells can be so limited in many species. This means that although in due course we might expect to understand the mechanisms of translocation in a few species, we should not extrapolate and predict the same mechanisms for all plants. *Laxmannia* of the Liliaceae has very small phloem elements in the vascular bundles of the narrow leaves; these elements are embedded in a matrix of fibres and would appear to have very indirect communication with the mesophyll.

The other area of applied study is in relation to diseases of the phloem. The normal structure of functioning phloem must be understood before disease symptoms can be interpreted, or the effects of chemical treatments defined.

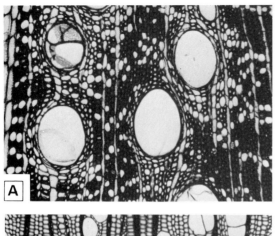

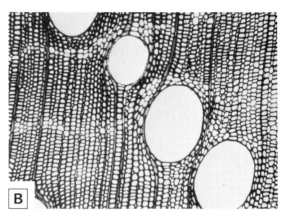

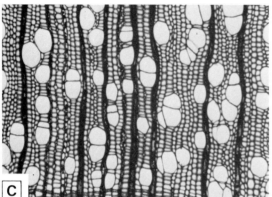

Fig. 6.12   Wood from members of the Fagaceae; all T.S., ×130. A, *Quercus brandisiana*; B, *Lithocarpus conocarpa*; C, *Nothofagus solandri*. A, like *Q. robur* (Fig. 6.8) has uniseriate and multiseriate rays although the illustration shows an area with only uniseriate rays. B and C have only uniseriate rays. A and B have tangential bands of axial parenchyma. Tyloses are present in A, and tracheids accompany the vessels in A and B. Fibres in A and C are strongly thickened. C has consistently narrower vessel elements than A or B, and has short radial vessel multiples. Fagaceae constitute a very natural family. There are two main anatomical groups as far as wood is concerned. A and B represent one group, C the other.

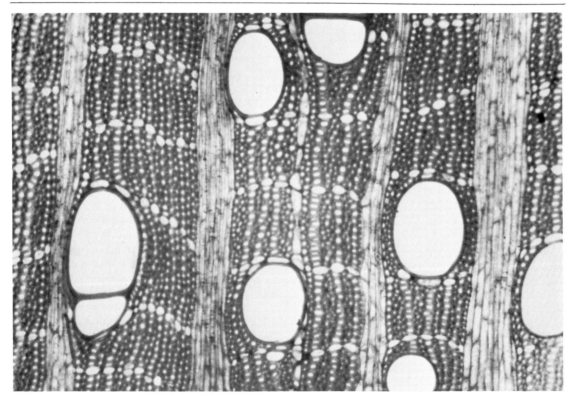

Fig. 6.13   *Platymitra siamensis*, Annonaceae. Vessels diffuse, porous; rays uniseriate and multiseriate; axial parenchyma in uniseriate tangential bands; fibres thick-walled. T.S., ×130.

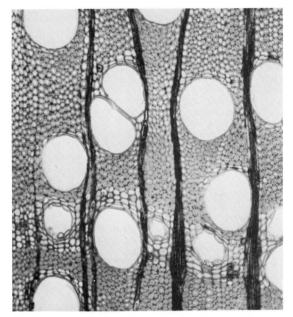

Fig. 6.14 *Hopea latifolia*, Dipterocarpaceae. Vessels diffuse, porous – solitary or in pairs. Rays mainly multiseriate, widening at parenchyma bands. Axial parenchyma in broad tangential bands. Conspicuous axial secretory canals are present in nearly all members of the family. T.S., × 130.

Some softwoods in which particular features can be found

Axial parenchyma *Sequoia, Taxodium.*
Bars of Sanio *Sequoia sempervirens.*
Pitting (of tracheid to tracheid walls):
    alternate biseriate *Agathis palmerstonii*
    multiseriate *Taxodium distichum*
    multiseriate alternate *Araucaria angustifolia*
    opposite biseriate *Sequoia sempervirens*
Rays, tall, *c.* 30 cells *Abies alba*
Rays, low, most less than 10 cells *Juniperus*
Ray tracheids *Pinus, Picea, Larix*
Resin ducts, axial *Pinus, Picea*
Resin ducts, radial *Picea, Pseudotsuga*
Spiral thickening *Taxus, Juniperus*
Torus margin, irregular *Tsuga heterophylla*
Torus margin, scolloped *Cedrus*
Window pits *Pinus sylvestris*

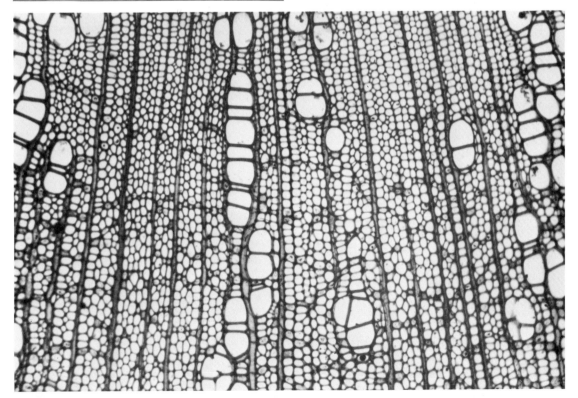

Fig. 6.15 *Carpinus betulus*, Carpinaceae. Vessels diffuse, porous, in long radial multiples. Rays uniseriate (aggregate rays also occur, but are not shown). Axial parenchyma can be seen in poorly defined, interrupted tangential bands, the cells have dark contents. T.S., × 130.

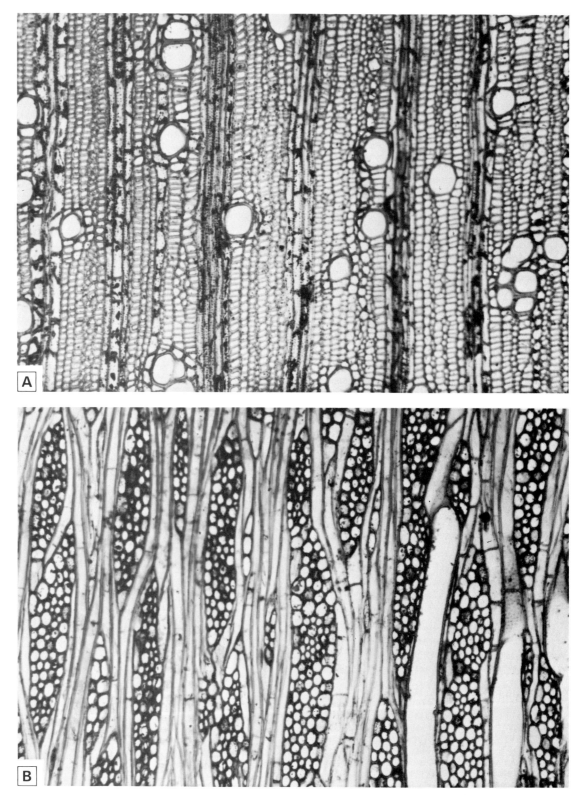

Fig. 6.16 *Laurus nobilis*, Lauraceae. A, T.S. B, T.L.S. Vessels diffuse porous, narrow, solitary or in small multiples, perforation plates simple. Rays uniseriate and multiseriate, heterocellular. Fibres septate., × 130.

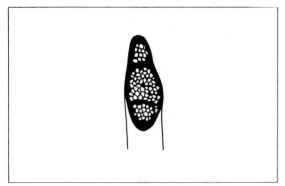

Fig. 6.17　*Aesculus pavia*, compound sieve plate, × 720.

Some characters in the secondary xylem of selected hardwoods

The descriptions given here supplement the main text, and give examples of some of the features mentioned there. They are not intended to be complete.

Azadirachta indica Meliaceae Vessels solitary and in radial multiples, perforation plates simple, intervascular pitting fine; rays 1–4 cells wide, heterocellular; parenchyma vasicentric and in narrow tangential bands; crystals rhombic, chambered, abundant; gum in some vessels.

Buxus sempervirens Buxaceae Vessels narrow, mostly solitary, perforation plates scalariform with many bars, oblique; rays 1–2 cells wide, heterocellular, marginal cells upright, central cells procumbent; parenchyma diffuse.

Ceiba pentandra Bombacaceae Vessels mostly solitary, perforation plates simple; rays up to about 8–15 cells wide, heterocellular; parenchyma vasicentric and in narrow tangential bands alternating with narrow bands of fibres; tannin or resin in many cells, crystals present.

Dipterocarpus alatus Dipterocarpaceae Vessels wide, mostly solitary, perforation plates simple, transverse; tyloses present; rays 1–4 or 5 cells wide, heterocellular; parenchyma vasicentric and apotracheal, scattered and in tangential bands; fibres thick-walled; vertical canals with thin-walled epithelial cells set in broad bands of tangential parenchyma.

Dombeya mastersii Sterculiaceae Vessels solitary and in short radial multiples, perforation plates simple, intervascular pitting alternate, pits circular; rays 1–4 cells wide, heterocellular; parenchyma aliform to aliform confluent; fibres thick-walled.

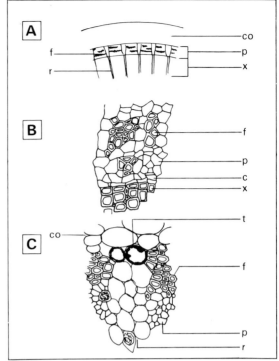

Fig. 6.18　A, diagram to show location of phloem fibres in *Tilia* stem T.S. B, *Malvaviscus arboreus*; C, *Gossypium* sp., both × 218. c, cambium; co, cortex; f, phloem fibres; p, functional phloem; r, ray; t, tannin; x, xylem.

Eucalyptus marginata Myrtaceae Vessels solitary and in radial and oblique multiples, perforation plates simple, tyloses present; rays 1–2 cells wide, heterocellular; parenchyma mostly amphivasal; fibres thick-walled, dense, septate.

Liriodendron tulipifera Magnoliaceae Vessels wide, thin walled, in radial, tangential and oblique multiples, occupying most of volume of wood, perforation plates scalariform, oblique, intervascular pits wide, opposite; tyloses present; rays mostly 2–3 cells wide, expanded at growth rings, heterocellular.

Pistacia lentiscus Anacardiaceae Vessels in long, radial multiples, some elements much wider than others, perforation plates simple; tyloses present; rays mostly 1–2 seriate, heterocellular, some with secretory canals; some fibres septate.

Pittosporum rhombifolium Pittosporaceae Vessels rounded-angular, solitary or in radial to oblique groups, perforation plates simple, vessels with tails and very fine spirals; rays mainly 3–4 cells wide, heterocellular, some only one cell.

Rhododendron sp. Ericaceae Vessels angular, solitary or in small groups, perforation plates scalariform, with many bars, some spirals present; rays 1–4 cells wide, some only one cell.

Robinia pseudacacia Papilionaceae Ring porous; wide vessels solitary or in short radial multiples, narrow vessels in clusters, perforation plates simple, transverse, intervascular pits vestured; tyloses present; rays storied, most 4–5 cells wide, more or less homocellular; parenchyma aliform confluent, storied.

Sparmannia africana Bignoniaceae Vessels angular, solitary or in short multiples, or in clusters, perforation plates transverse, simple, intervascular pits large, with narrow borders; rays 1 to 8 or more cells wide, composed of wide cells, heterocellular; rays making up large proportion of wood; fibres sparse.

Tectona grandis Verbenaceae Growth rings conspicuous; vessels solitary, in pairs or radial multiples, perforation plates simple, intervascular pitting fine, alternate; tyloses; rays mostly 1–3 cells wide, heterocellular; parenchyma initial and a little vasicentric; fibres septate; deposits in some vessels.

Further Reading

Brazier, J. D. & Franklin, G. L., 1961. *Identification of Hardwoods. A Microscopic Key*, Forest Products Research Bulletin, No. 46, H.M.S.O., London.

British Standards 881 and 589, 1974. *Nomenclature of Commercial Timbers, Including Sources of Supply*, British Standards Institution.

Jane, F. W., revised by Wilson, K. & White, D. J. B., 1970. *The Structure of Wood*, 2nd edn, A & C Black, London.

Kribs, D. A., 1959. *Commercial Foreign Woods on the American Market*. Pennsylvania State University Press.

Phillips, E. W. J., 1948. *The Identification of Coniferous Woods by their Microscopic Structure*, Forest Products Research Bulletin, No. 22, H.M.S.O., London.

See also list of general advanced plant anatomy texts, p. 14.

7 Adaptive features

The relationship between plant structure and the environment in which the plant grows was a subject that fascinated the early plant anatomists, and continues to be of great interest today.

In the early history of such studies, correlations were made on a rather empirical basis. Plants found growing in, for instance, dry conditions were studied and shown to have anatomical modifications not normally associated with plants from more mesic localities. Without any attempt at experimentation, the authors of that period would ascribe specific properties to the structures they saw. For example, Haberlandt's book, *Physiological Plant Anatomy*, was written largely from observation, and must therefore be used with care. Many researchers subsequent to Haberlandt have adopted his ideas uncritically. Where people have taken the trouble to study the anatomy of a range of plants from one habitat, they have found some features which seem to vary so widely in expression – for example thickness of the walls of epidermal cells – that their adaptive significance is put in doubt. There are, however, certain types of modification which crop up with such a degree of regularity, and in such taxonomically diverse plants, that they might really be related to survival in that particular habitat.

Despite any adaptations found in the anatomy of plants which might be thought to be of 'ecological' benefit, it is normal for the family or genus characters to be well expressed and often dominant.

Not all adaptations are evident at the morphological level. Some are physiological, and physiological races of plants have evolved which fit them for growth in extreme conditions. For example, some races of *Agrostis* species can grow in areas of high concentration of heavy metals, e.g. copper, where other plants fail. These adapted grasses have been shown to accumulate and immobilize heavy metals in their roots.

The duration of life of the plant might be a dominant feature which helps a species to survive. Ephemeral species may grow in normally xeric conditions if they can germinate, grow, flower and fruit when water is available. During this short period of activity, the plant may have adequate water and would not need any other xeromorphic adaptations.

The issue is made more complex when it is realized that there are often many microecological niches even within a small area. Diversity in anatomy could relate to such differences which are often hard to detect without prolonged study of the area concerned. Seasonal variability in environment may be overlooked by those making plant collections at particular times of the year. It boils down to the observation that if a species is found growing successfully under a given set of conditions, it is there as a result of selection and adaptation, and ability to compete with other species for that niche.

Some of the main habitats and commonly associated plant modifications will be outlined below. Despite the cautionary remarks made above, it is often possible to find in plants anatomical features which do show a close correlation to the habitat type in which they normally occur.

Xerophytes

Xeric habitats. Plants growing under very dry conditions normally show a reduction in evaporating surface area. When leaves are developed they may be small, or have various features which would appear to assist them in regulating or reducing potential water loss. Leafless plants, e.g. many Cactaceae and Euphorbiaceae, and others with non-functional leaves, e.g. most Restionaceae, often have subspherical or more or or less cylindrical stems modified to perform the

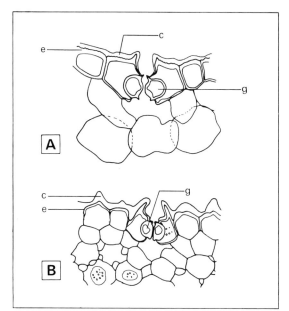

Fig. 7.1 A, *Aloe somaliensis*, outer part of leaf T.S.; B, *Haworthia greenii*, outer part of leaf T.S., both × 218. Note the sunken guard cells (g), the thick cuticle (c) and the thick outer wall to the epidermal cells (e). Both have succulent leaves, with little mechanical tissue.

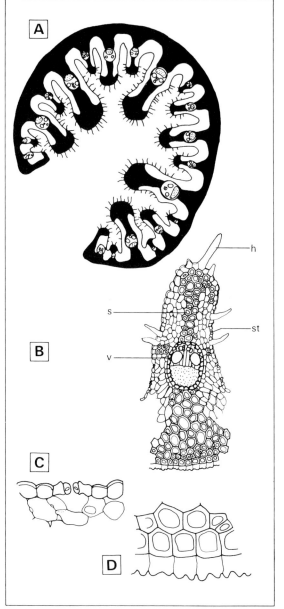

Fig. 7.2 *Ammophila arenaria* A, low power, plan leaf T.S., × 25. s, sclerenchyma; v, vascular bundles; h, hairs; st, stoma. B, detail of rib (black areas represent thick walled cells), × 54. C, adaxial epidermis with stoma. D, abaxial epidermis with very thick cuticle. C and D, × 300.

photosynthetic and transpirational functions normally ascribed to leaves.

A sphere has the smallest surface area possible for a given volume. Cylinders also have a low ratio of surface area to volume.

The bulbous habit is often related to dry situations; flowers and leaves are present as aerial organs for a limited period each year, e.g. *Narcissus, Tulipa, Haemanthus, Scilla*. Swollen underground stems also occur, e.g. many Asclepiadaceae, rhizomes, e.g. *Iris* species, or corms, e.g. *Crocus, Watsonia*. As in ephemerals, these plants normally grow actively when water is available, and their leaves in consequence may show little adaptation to xeric conditions.

In those plants which have persistent (perennial) leaves or stems, morphological and anatomical modifications are quite common. Stomata are often (but not always) sunken; they can be provided with various ante-chambers and cutin lined substomatal cavities which could play a part in water regulation. *Aloe* and *Haworthia* species (see Fig. 7.1) show some such modifications. The cuticle itself is often thicker in xerophytes than in mesophytes, but cuticle and epidermal cell wall thickness are not reliable guides to xeromorphy. Stomata may be very numerous and widely distributed, or they may be confined to grooves or channels in the leaf or stem. Some xeromorphic leaves are capable of inrolling (e.g. *Ammophila*, see

Fig. 7.2) and thus enclosing the stomata when dry conditions prevail. On the other hand, when adequate water is available, it has been shown that conifers with needle-like leaves can transpire as rapidly as mesophytes. Plants like *Aloe*, with thick cuticle, epicuticular waxes, thick outer walls to the epidermal cells, sunken and variously protected stomata seem to be well able to regulate and minimise water losses during dry periods. The raised rim forming a supra-stomatal cavity above each stoma may have a function enhancing evaporation when growth conditions are good. The structure could have a venturi effect, lowering pressure above the stoma and assisting transpiration.

Some plants, e.g. *Elegia*, can grow in areas of adequate ground water supply, but where strong drying winds could cause excessive water loss. These rush-like plants are not damaged physically by the strong winds since they are leafless and flexible. Many of the Restionaceae and some of the Juncaceae are remarkable in showing xeromorphic features in the stems but hydromorphic features in the roots. There is abundant mechanical tissue, usually sclerenchyma or other lignified cells, in the stems, but large air cavities in the cortex of the roots. It seems that the stems can be exposed to strong drying winds when it is probably too cold for the roots to deliver enough water to meet evaporation losses. The roots themselves often grow in waterlogged soils or standing water. Consequently, when conditions for transpiration and root action are satisfactory, the root adaptation is probably beneficial.

Internal adaptations in xerophytes may take one of two main forms; they can be for water storage, and the plants are then described as being succulent, or they can be for structural rigidity, with the ability to resist collapse and tearing on drying, and then the plants might be described as sclerotic. Tearing and disruption of tissues is one of the main causes of permanent injury resulting from excessive desiccation. Chlorenchyma enclosed in rigid, lined channels is less likely to be torn than that which is in an unprotected, relatively unstiffened mesophyte leaf. In succulent plants there is very little, if any, mechanical tissue, and the xylem of the vascular system is usually not strongly thickened. Many Crassulaceae (Fig. 7.3), Aloes etc. are of the succulent type and *Hakea*, *Leptocarpus* (Fig. 7.4) and *Ulex* are of the sclerotic type, as is *Ecdeiocolea* (Fig. 7.5).

Some *Haworthia* and *Lithops* species show only translucent leaf tips above the ground level. The rest of the leaf is buried, but contains chlorenchyma and water storing mesophyll. These are often called 'window' plants. Light is able to penetrate to the photosynthetic tissue through the translucent cells.

Some other characters often associated with xerophytes are of a much more dubious nature. Hairs have been supposed to help reduce surface wind speed and hence evaporation rates, but hairiness is often much more of a family character and there are

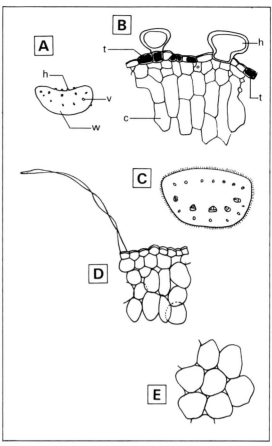

Fig. 7.3 A, B, *Crassula* sp. C–E, *Senecio scaposus*. A and C, plan T.S. leaf; mechanical tissue absent, central mesophyll cells store water. B, detail of outer part of A. D, outer part of C. E, central part of C. B, D, E, ×54. c, chlorenchyma; h, hair; t, tannin; v, vascular bundle; w, water storage tissue.

many xerophytes belonging to families in which hairs are rare that manage quite adequately without them. Thin-walled hairs could easily increase water loss under some conditions, but most hairy xerophytes, e.g. *Gahnia* spp., *Ammophila* and *Erica* spp., have hairs with thickened walls and some also have thick cuticle. There are numerous mesophytes with hairs.

Many xerophytes have a hypodermis, the cells of which are thick-walled (Fig. 7.5).

Compact mesophyll, with few air spaces has also been thought of as a xeromorphic character, e.g. *Pinus* spp. with plicate mesophyll cells, however, many succulent and sclerotic xerophytes have mesophyll or stem chlorenchyma that has abundant air spaces, e.g. *Laxmannia*, *Hypolaena*. Experiments are needed to determine the significance of high volumes of internal atmosphere which can be a feature of both xeromorphic and hydromorphic plants. Could the

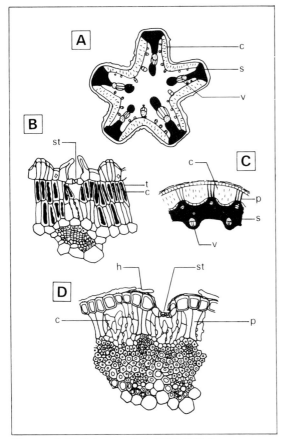

Fig. 7.4 A, B, *Hakea scoparia*, leaf T.S. C, D, *Leptocarpus tenax*, stem T.S. A and C, ×15. B and D, ×120. Note sunken stomata (st) in both and abundant strengthening sclerenchyma (s). Hairs (h) cover the *Leptocarpus* and tannin (t) is present in the chlorenchyma of *Hakea*. The pillar cells (p) in *Leptocarpus* divide the chlorenchyma into longitudinal channels. c, chlorenchyma; v, vascular bundle.

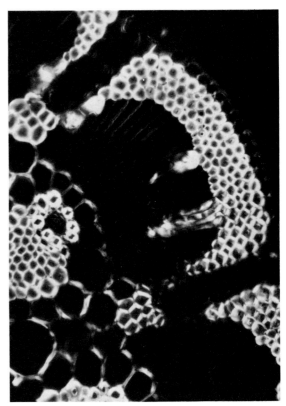

Fig. 7.5 *Ecdeiocolea*, outer part of stem, T.S. in polarized light, ×550. Fibres and sclereids show as lighter cells. This xerophyte has a deeply grooved stem, with stomata on the flanks of the grooves. Note the strong development of hypodermal fibres to the outer side of the thin-walled chlorenchyma.

function be the same in both types of plant, since they could each be growing under conditions which permit only a low rate of flow in the transpiration stream?

There are certain montane parts of the world where little, if any, surface water can be seen for many months of the year, and where soil water is often frozen. The cushion plant is the characteristic life form in such places. From observation of the compact habit, reduced leaf surface area, short internodes, widely penetrating roots and slow growth rate, it might be thought that the anatomy of all such plants would conform to a xeromorphic type. However, this has not proved to be entirely true. Some species, such as *Pycnophyllum molle* and *P. micronatum* of the Caryophyllaceae do show the expected adaptation. They have extreme reduction of the leaf surface area, leaves are closely adpressed to the cylindrical stems,

and sunken stomata are present only on the adaxial, protected surface, amongst papillae. But other species, like *Oxalis exidua* (Oxalidaceae) have very little apparent xeromorphic modification. *O. exidua* has hairs and papillae on the leaves, but the stomata are superficial. The chlorenchyma of the leaves is not compact. Its leaves are similar to those of the mesic members of the genus. The stem and roots exhibit an interesting modification, probably useful to a plant which has to penetrate the cracks between rocks. There is no interfascicular xylem or phloem produced during secondary growth and the vascular bundles remain separate. Cambium in the interfascicular regions produces parenchyma. Apparently, the roots and stems can twist and deform without undue compression of the vascular supply, rather as in many lianes.

Azorella compacta (Umbelliferae) does show anatomical features which appear to be related to the harsh high montane environment. The leaves are small and very shiny (thus reflecting ultraviolet light).

Contractile roots are present which help to keep the plant firmly anchored, despite frost heave. The vascular bundles are separate, as in *Oxalis exigua*. Resin ducts, a family characteristic, are frequent.

Anthobryum triandrum (Frankeniaceae) is also well adapted to cold, drying conditions. The leaves are furrowed, with stomata confined to the furrows. However, the vascular cylinder is compact, not composed of separate bundles.

So, once more we have evidence that some 'family' characters can be conserved in much modified and reduced plants, and that various species with diverse anatomy can cope with a particular set of environmental conditions.

Halophytes often show succulence which is normally associated with dry conditions; they grow in saline areas where there is, in effect, a physiological drought. Although surrounded by water, the roots have to extract it from the soil against a considerable suction force. (This point has application in liquid feeding of plants through a gravel bed in glasshouses. The gravel has to be periodically flushed with salt-free water or the salt concentration becomes high enough to dehydrate the plants.)

Plants growing in soils which freeze for part of the year are also subjected to 'drought' conditions. Many conifers in such habitats have needle-like leaves and the ability to regulate water flow adequately, both in conditions of adequate and inadequate water supply.

The sap of plants which commonly flower and produce leaves before the snow and frost have gone is often of a mucilaginous nature, and acts as a kind of antifreeze.

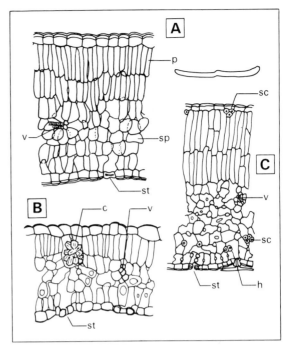

Fig. 7.6 Small parts of mesophyte leaves (lamina) in T.S. A., *Arbutus unedo*, × 109, B, *Corylus avellana*, × 120, C, *Olea europaea*, × 109. c, cluster crystal; h, hair; p, palisade; sc, sclereid; sp, spongy mesophyll; st, stoma; v, vascular bundle.

Mesophytes

Mesic conditions are suitable for broad-leaved plants with fairly soft, thin or somewhat sclerified coriaceous leaves. In temperate or tropical sub-montane zones, many mesophytes pass the winter months in a leafless form, either as deciduous trees or perennial herbs, and their buds have bud scales. The fringes of the mesic zones tend to have a higher proportion of evergreens with coriaceous leaves. There is a gradation from mesic to xeric conditions in many areas, and plants showing adaptations to both situations may grow side by side. Mesophytes tend to have anatomical variations which are related more to the family from which they come than to the environment in which they grow.

Consequently, it is difficult to generalize about the anatomy of mesophytes. The epidermal cells frequently have only moderately thickened outer walls and a thin or slightly thickened cuticle. The stomata, normally confined to the lower surface, are usually superficial. The mesophyll normally consists of one, two or more layers of closely packed palisade-like cells. Cells of the inner layers may be the least densely packed, and border the loosely arranged spongy mesophyll. Sclerenchymatous tissue is absent or sparse, and may be represented by a small number of sclereids. Sclerenchymatous sheaths to vascular bundles are rare, except in relation to larger primary veins, midrib or petiole. See Fig. 7.6 for examples of mesophyte leaves.

Tropical rain forests are adequately supplied with water. The dominant life form is the very tall tree. Leaves are often long-lived, because there are few seasonal stresses which might necessitate the adoption of a regular deciduous habit. Some rain forest trees continuously lose and replace leaves. Others may be deciduous every few years; these often flower before the new leaf growth. The leaves are often relatively tough (coriaceous) but have large surface areas.

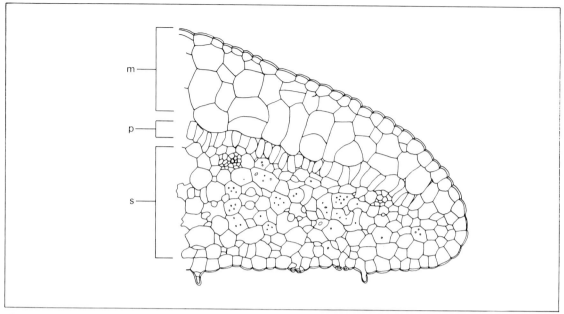

Fig. 7.7 *Codonanthe* sp. Part of leaf T.S. showing multiple epidermis, m, a single palisade, p, and a large quantity of spongy mesophyll, s, × 102.

Many have an extended 'drip-tip'. Bud scales are rarely developed.

The relative humidity inside the canopy of the rain forest is normally approaching 100 per cent. Many epiphytes grow in the relative shade. Some, e.g. certain Bromeliaceae, may have a construction by which leaves funnel water to the centre of the plant where it is held in 'tanks' formed by leaf bases. The normal roots of these and other epiphytes are simply anchors, and do not extract nourishment from the plants upon which they grow. Many of the epiphytic Araceae and Orchidaceae have special aerial roots with modified enlarged epidermal and cortical tissues (velamen) which can absorb and retain atmospheric moisture (Fig. 4.32). In the shelter of the large trees, understories of shorter trees flourish, often with frond-like leaves of great length.

A number of the epiphytic gesneriads in the Old World tropics have a particularly interesting and as yet unexplained adaptation. The upper epidermis of the leaves is multi-layered, and consists of colourless cells. In some species it may make up to two thirds of the total thickness of the leaf. The chlorenchyma is relatively thin, with a conspicuous layer of widely spaced palisade-like cells and some spongy mesophyll (Fig. 7.7).

Many of the epiphytic Bromeliaceae have numerous hairs and scales on their leaves, thought to be capable of absorbing water from the very humid atmosphere in which the plants grow.

Apart from the root modifications and the somewhat coriaceous leaves of the upper storey trees, rain forest tracheophytes appear to have more anatomical characters relating to the families to which they belong than to the environment in which they grow.

Hydrophytes

Hydrophytes, plants growing immersed in water or with the leaves floating and, perhaps, aerial inflorescences, show many anatomical features which clearly relate to their habitat, and in some instances family characters are so reduced as to be difficult to define.

Most stems and leaves have large air spaces between layers of internal tissues. These assist in buoyancy and also gas exchange. Cuticle is poorly developed or absent. Stomata are usually absent from submerged surfaces, but may be present on the upper surface of floating leaves. Vascular tissue, particularly xylem, is poorly developed and sclerenchyma is normally absent (Fig. 7.8).

Morphological adaptations include the reduction or absence of lamina or very linear leaf form in the submerged leaves of plants growing in running or tidal water, e.g. *Zostera, Posidonia*.

Plants growing in acid bogs have particular problems to overcome, particularly since mineral concentration is low in the water, and nitrogenous salts are

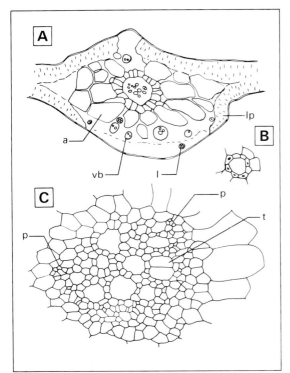

Fig. 7.8 *Limnophyton obtusifolium*, part of midrib, T.S. A, diagram, × 15, showing large air spaces around central vascular complex. B, laticifer, × 110. C, central vascular tissue, × 200. a, air space; l, laticifer; lp, loose palisade chlorenchyma; p, phloem; t, tracheary element.

almost absent. A number of plants from different families have developed anatomical features which help them to survive such conditions. Among these, the animal trapping (so-called 'insectivorous') plants are of particular interest. All have specialized glandular hairs on the leaf surface, e.g. *Pinguicula, Drosera*. These hairs may be of two types, stalked ones secreting very sticky substances which trap the victim, and the other, sessile, secreting digestive enzymes. The leaf gradually rolls over to enclose the animal and opens out again when digestion and absorption is complete. Another, *Dionaea*, has sensitive trigger hairs on the lamina, three on either side of the midrib. The hairs are hinged at the base. They require two or three tactile stimuli to cause the leaf to fold shut vigorously. Marginal teeth mesh together, forming a prison from which the prey cannot escape. Reddish,

glandular hairs then secrete digestive enzymes and absorption follows. Specialized hinge or motor cells are present along the midrib.

Applications

The application of information about anatomical modification in plants developed in response to various environments may at first sight seem obscure.

The morphology and anatomy of a plant can give the horticulturalist a good guide as to the sort of growing conditions he should provide. Take for example an orchid with conspicuously swollen leaf bases, indicating the facility for water storage, and with aerial roots. It would be clear that the plant was an epiphyte which required support on a branch or log, and that it would need a very humid warm atmosphere. On the other hand, a plant with a rosette of thick, succulent, closely packed leaves with transparent tips would clearly be xeromorphic, would need to be planted deep in a quickly draining soil or compost so that the leaf tips were level with the surface. It would need bright light and probably a period of each year with little or no watering. It would need to be protected from frost, and may need additional heat.

The taxonomist also finds anatomical data of importance when dealing with plants from different families that have made a parallel response to a given environment, producing a similar morphology. A number of the monocotyledon families are like this. Since it is quite common for the anatomy to have retained some characters which are diagnostic for the family, these can be applied to solve problems of affinity. Both xerophytes and some hydrophytes are amenable to this type of study.

The plant breeder might find it worth looking at the anatomy of wild relatives of crop plants if he wishes to integrate some drought resistance, or extra structural rigidity, for example, into the crop.

It is clear, then, that some of the anatomical features which are apparent in plants are modified to an extent in relation to the environment in which the plants grow. No generalizations or sweeping statements can be made, though, and each species must be assessed on its own merits.

8 Flower and fruit

Apart from their ornamental or horticultural value, flowers have been studied mainly as a source of very important taxonomic characters and in relation to phylogeny and evolution. Their prime function in reproduction has naturally been the object of vast amounts of morphological and physiological investigation.

The extreme importance of fruits and seeds as food has provided the inspiration for a great deal of research.

However, in this volume which deals largely with vegetative anatomy, only matters of particular interest relating to flower and fruit can be outlined. The reading list at the end of the chapter will assist those who wish to learn more on the subjects mentioned here.

Vascularization

Many of the anatomical features used in comparative studies are to be found in the arrangement and number of vascular bundles and their types of branching in inflorescences, flowers and floral parts. These patterns can be hard to interpret. Are all or most of the bundle pathways genetically predetermined? Could a significant proportion of bundles be present as responses to physiological demand, that is, related to physiological needs which are to be met rather than some archaic pattern recapitulating ancestral conditions? Despite the difficulties of observation and interpretation, many valuable studies have provided data on vascularization which enable us to understand the interrelationships between many genera and families of flowering plants.

Those whose interests lie in the phylogeny of flowering plants, or in the origins of angiosperm flowers, make considerable use of the results of studies on vascularization. It is widely held that vascular systems in flowers are conservative, that is, they may remain relatively unchanged even when the general shape of the flower has altered by evolution. This may lead to the formation of odd-looking loops or curves in some of the vascular strands to accommodate changes in the relative positions of the floral parts. In some flowers, small branches of the vascular system end blindly. This can be taken to mean that in one or more of the ancestors of the plant similar strands served some organs or appendages which are lacking in the present day representative. For example, a modern, unisexual female flower might have remnants of a vascular system which would have served stamens in a bisexual ancestor.

When organs are adnate, for example, a stamen fused with a petal, it often follows that the vascular supplies of the two become fused into one strand. Bundle fusions can make it more difficult to interpret the vascular systems of flowers in comparative studies.

The number of traces to each floral organ can vary. Often petals have only one trace, but petals in certain families regularly have three petal traces. The number of traces to each sepal is often the same as that to the foliage leaves of the same plant. Stamens may have one or three traces, but one is by far the most common number. Carpels may possess one, three, five or more traces. Dorsal and marginal or ventral traces are distinguished in descriptions, when three or more are present.

Sometimes a flower may have very unusual morphology which is difficult to interpret. It could be that examination of its vasculature would help one to understand the true nature of the various parts. If other members of the same genus or family have more normal flowers, then comparative studies could prove most informative.

A number of theories concerning the origin of the angiosperm flower have their basis in comparative studies of floral and vegetative vascular patterns of both living and fossil plants. Despite the amount of work done by numerous people, there is no common consensus of opinion. Doubtless, new theories will be proposed. Some think that we have all the evidence we need, if only we will interpret it properly. Others consider that there are such large gaps in the fossil record that no one will ever be able to prove their theories!

Since the classifications of plants rightly give a great deal of prominence to characters of flower and fruits, there is a wealth of data on these parts. Consequently, it is normal to try to identify plants with flowers or fruit attached by reference to floras and herbarium specimens. Anatomical studies may help if the floral parts are in poor condition.

Palynology

Pollen studies have increased enormously with the advent of transmission and scanning electron microscopes. However, a great deal of foundation work was carried out with the light microscope; indeed a very sound basis was established. New tools have meant that details of fine surface patterning can be seen easily (Fig. 8.1). Surveys of families can now be carried out much more rapidly, and electron micrographs, particularly from the SEM are easy to interpret.

Pollen grains are often readily identified to the genus level, and sometimes to the species level, if adequate reference material is available. In some families there is great variability in pollen grain morphology and surface features; in others there is uniformity.

Besides the taxonomic inferences which can be drawn from studies in comparative palynology, the subject has a number of other applied aspects. For example, honey purity and origin can be determined by a study of the pollen grains it contains. Pure heather honey would not be expected to contain large quantities of *Eucalyptus* pollen! Adulteration can usually be detected microscopically.

Pollen grains shed from plants remain in a recognizable form in peat deposits for very long periods of time. By careful analysis of the pollen grains in successive layers of peats, or at successive levels, it is often possible to build up a picture of the vegetation of previous ages.

Pollen–stigma interactions

Most plants have mechanisms by which they can

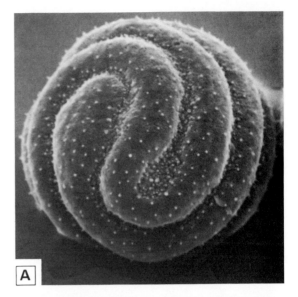

A

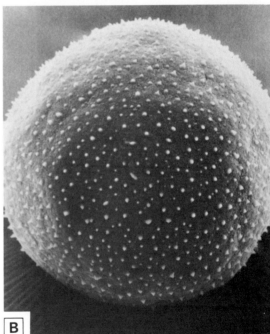

B

Fig. 8.1 Surface details of two pollen grains for comparison. A, *Crocus michelsonii*, B, *Crocus vallicola*, both SEM photographs, × 1,000.

'recognize' pollen grains from their own and other species. Stigmas often possess a complex of chemicals which enable them to respond to chemicals contained in the outer layers of pollen grains. The chemicals are usually proteins. Figure 8.2 shows a pollen grain germinating on a stigma. In species which are not self-fertile, the stigma rejects pollen from anthers pro-

p

s   pt

Fig. 8.2 *Tradescantia pallida*, pollen grain germinating on stigma. p, pollen grain; pt, pollen tube; s, papilla on stigma. Freeze dried, viewed in SEM, × 1,000.

duced by the same or other flowers on the plant. Many plants normally reject pollen from other species. However, some species are inter-fertile. In nature they might not normally be pollinated by other species. Their flowering periods might not coincide, or they may be too distant from one another. In some plants the stamens are mature well before the stigma (protandry) and pollen is dispersed before the stigma is receptive. In others the stigma may mature and senesce before pollen from the same flower is released (protogyny).

Pollen can be stored alive at low temperatures, so we can try out crosses even if flowering periods are different. The horticulturalist attempting to maintain pure species, is only too aware of the crossing which can go on between related plants brought together in one glasshouse.

Some of the rejection mechanisms result in obvious physical changes in the stigma. For example, reactions may occur which cause the stigma to callus over, so that the pollen tube cannot enter. Sometimes the size of the stigmatic papillae may be too great for small pollen grains to germinate on them successfully, or they may be too small for large grains, and also prevent effective pollination. Although pollen from closely related species may be accepted by a stigma this is not always the case. Incompatibilities may arise. In another mechanism which prevents different species from crossing, the length of the style may be much longer than the potential length of the pollen tube, so that fertilization cannot take place.

Where it is desirable horticulturally or agriculturally to produce hybrid plants, anatomical and histochemical studies of the pollen–stigma interactions can help us to manipulate the process and override the blocking mechanism.

The system which promotes outcrossing may simply rely on different relative heights of anthers and stigma in the flower, as in pin-eyed and thrum-eyed Primulas. In the first, the stamens are short and the stigma is elevated on a long style; in the second anthers are at the outer end of the corolla tube, and a short style keeps the stigma at a low level. Insects visiting a pin-eyed flower are more likely to deposit pollen on the stigma of a thrum-eyed than of a pin-eyed flower.

Pollen itself can sometimes be cultured, and made to produce haploid plants.

Embryology

Embryo studies fall into two categories, first the comparative and developmental studies and second those aimed at embryo culture (or haploid, embryo sac culture).

Embryology and the sequence of cell divisions involved in embryo sac formation and following fertilization have become very specialized fields of study. There is a large body of comparative data available to the student; most of which is applied to evolutionary and taxonomic studies.

Embryo culture involves dissecting out embryos and growing them in culture media. This is sometimes done to ensure development and establishment of particular individual plants, but more often as a means of vegetative propagation.

Seed and fruit histology

The wide use of seeds and fruits in human food and animal feedstuffs has made knowledge of their anatomy of paramount importance. It is essential to be able to identify fragments of seeds and fruits in relation to possible adulteration and purity.

Although many species have been studied for seed

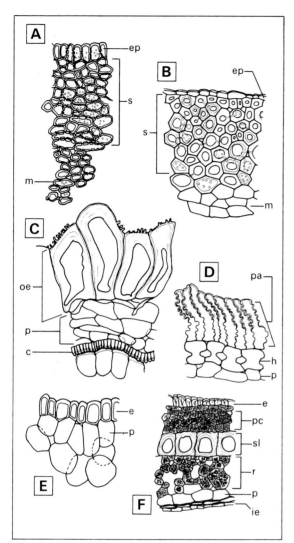

Fig. 8.3 Fruit wall and seed coat details in T.S. A, *Aesculus hippocastanum*, outer part of fruit wall, × 109. B, *Fagus sylvatica*, outer part of fruit wall, × 109. C, outer part of seed coat of *Delphinium staphisagria*, × 109, note small outgrowths from epidermal cell walls. D, *Cicer arietinum*, seed coat, × 218. E, *Cola acuminata*, seed coat, × 218. F, *Cucurbita pepo*, × 109. c, cells with U-shaped wall thickening; e, epidermis; ep, epicarp; h, hour glass cell; ie, inner epidermis; m, mesocarp; oe, outer epidermis; p, parenchyma; pa, palisade cells; pc, pitted cells; r, reticulate spongy parenchyma; s, sclerenchyma; sl, sclerenchyma layer.

importance have received most attention. The main cereals, oil seeds and edible leguminous seeds have been described anatomically, as have those of selected weeds and poisonous plants. Fig. 8.3 shows some fruit walls and seed coats in T.S. Good sources of information are the specialist books and reference books on the anatomy of food plants.

Enough is known for us to realize that good comparative studies of seed coat can yield taxonomic characters of some significance. In families like the Umbelliferae, fruit anatomy provides many useful diagnostic and taxonomic characters. Undoubtedly, as more families are systematically studied, much of taxonomic and perhaps phylogenetic importance will emerge.

As with pollen studies, interest in seed coat anatomy has been stimulated by the general availability of scanning electron microscopes. It is now often possible to detect minute differences in seed coat patterns which might enable us to define species characteristics. Seeds do tend to vary a lot in size and sometimes in shape within a species. Their surface patterns go through developmental stages and consequently only mature seeds should be studied for comparative purposes.

The adaptive or physiological significance of surface features is by no means clear, and valuable studies on the development of patterns to maturity are of current interest. Changes which take place during germination are also studied, as are changes under storage conditions which could lead to deterioration of the seed and loss of viability.

The conditions required for germination are so specific and specialized for some seeds that elaborate experiments have to be conducted in order to discover them. Parallel anatomical studies can help in the interpretation of the results.

Further Reading

Davis, G. L., 1966. *Systematic Embryology of the Angiosperms*, John Wiley, New York.

Martin, A. C. & Barkley, W. D., 1961. *Seed Identification Manual*, University of California Press.

Vaughan, J. G., 1970. *The Structure and Utilization of Oil Seeds*, Chapman & Hall, London.

and fruit structure, relatively few families have been carefully studied systematically and in detail, followed by documentation of the results. Plants of economic

9 Economic aspects of applied plant anatomy

Many of the applied aspects of plant anatomy have been referred to in the previous chapters, but some do not fit well into descriptive text. I have therefore amplified some of these examples in this chapter and introduced new ones drawn from my experience at the Jodrell Laboratory, Kew. In writing this chapter, I have had to be very selective – a whole book could be written on this subject alone – but I hope that the following will serve to show the wide range of applications to which a knowledge of plant anatomy can be put.

Identification and classification

It is not always appreciated how important it is to be able to give the correct name to a plant. Cytologists, geneticists, plant breeders, chemists and anyone using plants for medicine, food, furniture, fabric or building material, or those conducting botanical research must be able to identify their source material or they may not be able to continue with their work. They would not know if further plant specimens or timbers were from the same species that they started with; their results and applications would be unpredictable; the foundations for scientific botanical research would be undermined. Identification depends on a stable, logical, usable and basically sound system of classification. At present, many plants can be identified adequately if all organs, e.g. flowers, fruits, leaves etc., are present. The traditional herbarium methods can then be applied. However, there are very large numbers of plants which have been classified using macromorphological features alone. A more natural, accurate and reliable classification results from also taking into account features of anatomy, palynology, biochemistry, population studies etc. This ideal can rarely be attained, but once the

'alpha' taxonomy of a family has been studied, the synthetic approach should be used for revisions, as has been done on the continent of Europe for a considerable time. I sometimes wonder if revisions based entirely on hand-lens studies of herbarium material should be outlawed!

Taxonomic application

Anatomical data are easily applied to improving classifications and can often be used in making identifications. Take for example the S.W. Australian genera *Anarthria* and *Ecdeiocolea*. Until recently these were treated as members of the Restionaceae. An extensive anatomical survey of the family Restionaceae showed the two genera to be bad misfits. Consultation with a classical taxonomist, Mr Airy-Shaw, proved that there were also taxonomically valid distinctions at the macromorphological level. Co-operative research resulted in two new families being recognized, Anarthriaceae and Ecdeiocoleaceae. Figure 9.1 summarizes some of the main differences. These families have subsequently been shown to be chemically distinct.

There are occasions when the herbarium botanist finds that it is difficult to ascribe a particular species or genus to a family, or where general affinities are suspected but there is insufficient evidence for him to place a taxon in a particular family. Here, additional anatomical evidence may be of help, but there seem to be about as many times when little extra helpful information comes from the anatomy. Recently, I examined tree leaf material collected in China; although there were no flowers or fruits on the herbarium sheet, the taxonomist, Mr L. Forman, thought that he knew the close relatives of the plant in question. The anatomy confirmed his views that the plant was *Pycnarrhena macrocarpa* Diels (Menispermaceae). A further study of species from this genus led to the discovery that two distinct genera

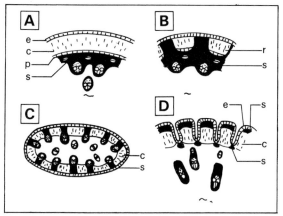

Fig. 9.1 Some differences between Restionaceae, Ecdeiocoleacea, and Anarthriaceae. A, B, Restionaceae. Stem T.S., most species have the general anatomy as shown in A, with a continuous parenchymatous sheath; in some genera the sheath is interrupted by extensions from the sclerenchyma cylinder, as in B. No vascular bundles occur in the chlorenchyma in all but 1 or 2 species. None of the species has hypodermal fibres or lacks a sclerenchyma cylinder as exhibited by Ecdeiocoleaceae, D. Anarthriaceae, C, also differ in having subepidermal fibre strands associated with vascular bundles; they may also have a sclerenchyma cylinder. Neither Anarthriaceae nor Ecdeiocoleacea has a parenchyma cylinder. c, chlorenchyma; e, epidermis; p, parenchyma cylinder (interrupted in B); s, sclerenchyma.

were involved and a new genus *Eleutharrhena* was named by Forman to include *P. macrocarpa* using evidence from morphology, anatomy and palynology. In *Pycnarrhena*, stomata are scattered over the abaxial leaf surface; in the *Eleutharrhena* the stomata are confined in clusters (Fig. 9.2).

Of course, the correct classification of plants is important, but it is often of more direct importance to know exactly to which species a specimen belongs. When flowers and fruits are absent, the plant anatomist comes into his own. Leaf fragments, wood and roots or twigs may have readily recognizable features which can be seen with a lens, but more often than not, identity has to be confirmed with the microscope.

Medicinal plants

Most of the drugs which we still extract from plants come from leaves, bark, roots or rhizomes. Leaves often become fragmented and detached; bark, roots or rhizomes can be difficult to identify from their macroscopic appearance. The proper authentication of crude drug material is essential for standards of safety and quality to be maintained. For these purposes, accurate anatomical and morphological descriptions of the drugs have been published. The legal standards are found in such volumes as the *British and European Pharmacopoeias* and the *British Pharmaceutical Codex*. In these books, the style of morphological and anatomical descriptions is very brief and to the point. Only those characters which will help to identify the material are given. Usually, these short monographs are carefully revised by a committee of experts at about five-year intervals. Herbalists are also aware of the need to have adequate control of the material they use, and work has been carried out to produce proper standards in a reference work.

It is quite often quicker to find out the identity of a crude drug from its anatomy than from its chemistry.

Importers of crude drugs are often experienced enough to know if they are buying pure material, or if adulterants are present. Sometimes samples will be sent for anatomical confirmation. For example Ipecacuanha, used in cough mixture can be adulterated with roots from alternative inferior species. Here microscopy can be used to give an indication of purity. The authentic source of the drug is *Cephaelis ipecacuanha* (Rubiaceae). Although rarely adulterated with other roots these days, there was a period when *Ionidium* (Violaceae) and other roots were regularly mixed in with the authentic material. Most of the adulterants have wide vessels in the xylem, whereas those in *Cephaelis* are narrow. The substitutes also lack characteristic starch granules, which are simple or, more usually, compound, with two or five or up to eight parts. The individual granules are oval, rounded or rounded and with one less curved facet; they rarely measure more than 15 µm in diameter. Sometimes *Cephaelis acuminata* is used as a substitute. This species is similar anatomically, but has starch granules up to 22 µm in diameter.

Sometimes closely similar substitutes are put on the market when the usual source of material is unavailable. For example, Bolivian Guarea bark is at present difficult to obtain, and a substitute from Haiti is available. Microscopic study has shown that the substitute is from a different species, since the groups of phloem fibres are dissimilar but chemical tests prove it to be equally suitable for use.

Occasionally the substitute may be poor and unsuitable. *Rheum officinale* root and rhizome is used medicinally, but *Rheum rhaponticum* is the vegetable. Fortunately, chemical and anatomical tests can be applied to detect which species is present.

Digitalis purpurea and *D. lanata* are used medicinally. They can be distinguished from one another on anatomical grounds, since the anticlinal walls of the abaxial epidermal cells are more beaded in *D purpurea*.

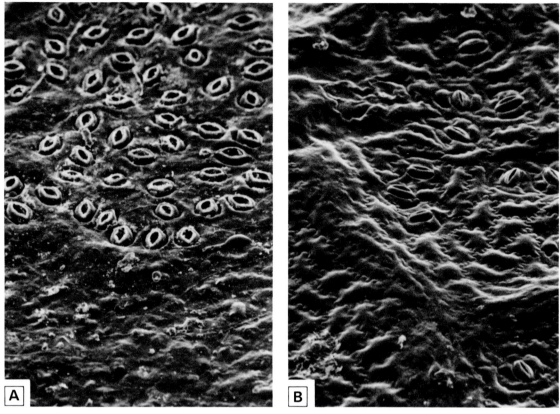

Fig. 9.2   Group of stomata in abaxial surface of *Eleutharrhena macrocarpa*, A. In B, *Pycnarrhena pleniflora*, the stomata are scattered over the abaxial leaf surface. Both SEM, × 300.

Herbal remedies used as folk medicines from tropical parts of the world are often only available in fragmentary form. Those wishing to determine the identity of such fragments need to use anatomical methods.

Food adulterants and contaminants

Some herbs are used extensively as seasoning. These are often imported in the form of dried powdered leaves. Again it would be easy to introduce useless or sometimes even poisonous adulterants which would be difficult to detect with the naked eye. We have examined samples of dried mint, *Mentha* species, for purity, only to find considerable quantities of *Corylus*, hazel leaf fragments included! *Ailanthus* leaf has also been used as a mint adulterant (see also p. 45, Fig. 4.25).

With the advent of the Trade Descriptions Act in the U.K., manufacturers must state the contents of their food products. It is essential for them to have

adequate quality control, and to be able to identify all the materials they use.

Foreign bodies sometimes get into food by accident. Often these are small and fragmentary and can be identified only with the microscope. A splinter of wood in butter was found to come from a species of *Pinus*. The importer and packers hoped to be able to determine if the splinter could have come from the country of origin of the butter, or whether it might have been introduced during the packing stages.

Buns and cakes containing sultanas periodically also contain other fruits which have become mixed with the sultanas during the sun drying process, when the sultanas are laid out in the sun. *Medicago* fruits are often involved. Some of these are prickly and unpleasant to eat!

We have examined an object from a tin of baked beans which looked remarkably like a piece of a mouse. It turned out to be a piece of rhizome from the parent plant. It is often the case that odd-looking inclusions in food are only pieces of the parent plants.

Vitis, grape vine stems have been found in currant buns, *Avena* coleoptile, looking like a mouse tail was present in a meat pie, and so on. Figure 9.3 shows an unsavoury looking shoot from a potato which occurred in a meat pie.

Starches from various plants have quite distinctive grain or granule features, so it is often possible to see if the stated materials have been used in a product unless the grains have become too hydrolysed (Fig. 9.4).

Animal feeds are made from the by-products of other food manufacturing processes, or from seeds and fruits grown specially for the purpose. When ground as a powder the constituents are difficult to detect by methods other than microscopy. There is plenty of scope for adulteration in feeds, and careful microscopical quality control is essential.

Animal feeding habits

Animal pests sometimes consume crop plants. It is often possible to find out what has been eaten by studying the composition of faeces, or stomach contents. A true estimate of potential losses can then be obtained. We have looked at faeces from rabbits, foxes, badgers, coypu, etc. and even millipedes! Of course the fragments of plant are very small when they have passed through an animal's digestive system. They are first fixed in FAA, then washed in water. Then comes a sorting process, using a binocular microscope. Similar looking fragments are put into a petri-dish, and the sample divided as far as possible into its components. Then fragments from each dish are examined using temporary mounts under the light microscope. We always hope for good characters like silica bodies, hairs, stomatal types and so on. It is a big help to have a set of reference slides made from vegetation growing in the area from which the animal concerned was captured.

It was suspected that some African cattle were being injured by eating grasses with sharp silica particles in them. The cattle only ate the grass concerned when other plants were unavailable. We examined the faeces and reported that there were silica bodies and sharp hairs present.

Domestic animals occasionally eat poisonous plants; and we may be called upon to identify the fragments. The owner of the animals can then take precautionary measures against further livestock poisoning.

Wood: present day

Most samples sent to Kew for anatomical identification consist wholly or mainly of wood. The samples

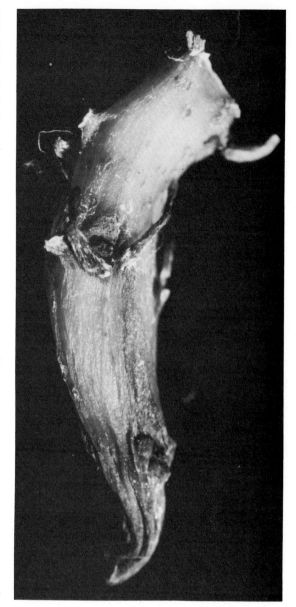

Fig. 9.3 Shoot of potato, from meat pie, mistaken for something worse!

are derived from many different sources and can be broadly divided into wood of recent origin and archaeological material.

Furniture is made from woods carefully selected for their appearance and strength. Fashions have changed and it is common for certain species to have been selected for a period and then superseded by others. In addition, some woods were unavailable at certain periods. Consequently, by knowing which species were involved in the manufacture of antique

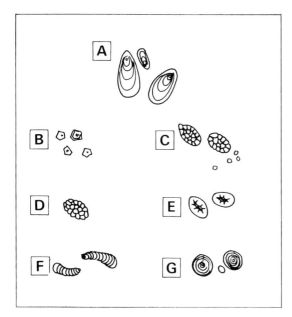

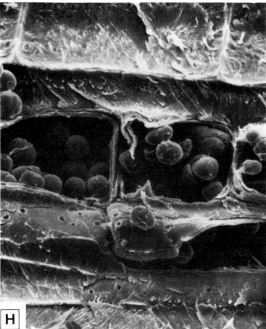

Fig. 9.4 Starch grains. A, potato; B, maize; C, oat; D, rice; E, pea; F, banana; G, wheat. All × 200. H, starch grains in xylem ray tissue of *Fabrisinapis*, SEM, × 3,000.

of being absolutely certain which woods were used is, in most instances, by making a microscopical study. Those who claim to be able to identify woods 'on sight' are either extremely experienced or over-bold, and many make errors.

The country of origin of carved wooden items can sometimes be established from the identity of the wood. Care must be taken because woods can be transported and then carved a long way away from their original sources. We have looked at items collected by Captain Cook on his voyages to try to determine where they could have come from, and this has proved successful. We once had for identification a wooden mask, carved in the likeness of a dog. This proved to be alder wood and its association with north American Indians was confirmed.

The Trade Descriptions Act has again provided problems for builders and manufacturers where woods are concerned. If they state that a particular wood has been used, this must be correct. The British Standards Institute has published a list of common names and the species from which the woods come, and this is the authoritative work which has to be followed. The only way to be certain that the correct wood has been used is to compare sections of it with those from a standard reference collection of microscopic slides. On one occasion a door said to be made of solid mahogany was brought to the laboratory. It turned out to be laminated, and no true mahogany was found in it – in fact the middle layer of veneer was birch!

Properties of woods related to structure have been mentioned in Chapter 6. We are occasionally asked to suggest substitute woods for some specialist purpose, when the supply of the normally used species has ceased. This can be difficult, but it is sometimes possible to suggest other species, which from their anatomical make-up might be expected to have similar properties.

Wood used as a backing for paintings, such as ikons, is brought to the laboratory from time to time. The purpose in finding out the identity is often related to establishing the name of the artist, or the country of origin.

In our time we have examined the wood from a good many walking sticks; an amazingly wide range of species has been used for this purpose!

Preservation of wood is of considerable economic importance. A great deal of experimental anatomy is carried out in various parts of the world in order to establish the nature of the process of decay, the identity of the organisms involved and the prevention of their degrading activities. The 'sound' wood has to be very carefully examined and described. Close

furniture, it may be possible to date the piece and occasionally the furniture expert may be able to get a good idea of who made it. Some craftsmen worked only with a carefully selected, characteristic range of woods. When repairs are necessary, it is also helpful to know which species should be used. The only way

observations then have to be recorded on all stages of the decay processes and the action that the various organisms have on the wood.

Considerable damage, running into millions of pounds, is caused each year to buildings either directly or indirectly by the action of roots of trees or shrubs. There may be a number of different tree species near to the buildings concerned. All or some of them might have roots beneath the foundations. It would be excessively expensive to try and trace the roots back to their parent trees by excavation. Fortunately it is possible to identify most roots of trees growing in the British Isles from aspects of their root anatomy, largely from features of the secondary xylem. In some instances it is possible to identify to the species level, but more often only the genus can be identified, for example *Quercus*, oak or *Fraxinus*, ash and *Acer*, *maples* and sycamore. In the Rosaceae, identifications can be made only to the subfamily level, e.g. Pomoideae and Prunoideae. Current research is aimed at finding additional characters in this family.

Sometimes it is not possible to get closer than the family, as for example in Salicaicae. In trunk wood, *Salix* and *Populus* can normally be separated because *Salix* usually has heterocellular rays and *Populus* homocellular rays. However, in the root wood this distinction does not hold.

Indeed, root wood is often slightly dissimilar in its anatomy from trunk wood of the same species. This means that one cannot rely on the descriptions contained in reference works on wood anatomy for accurate identification of roots. Root anatomy is also quite variable within a species, so the only way to be sure of making the proper identification is to compare the root sections with reference microscope slides taken from a range of authenticated specimens. Figure 9.5 shows two roots of *Acer pseudoplatanus* (T.S.) grown under very different conditions, and some normal trunk wood for comparison.

Wood: in archaeology

Wood, or charcoal is often preserved in sites from antiquity. The best preservation occurs in localities which are either very dry or continuously wet. Fluctuating drying and wetting encourages the activity of microorganisms and can lead to the rapid decay of wood.

Charcoal, usually in the form of fire ash or the burnt remains of structural posts in post holes, often retains even very delicate features of vessel element wall pitting and perforation plates. Figure 9.6 shows Romano-British *Alnus* charcoal. It can be difficult to see details of the anatomy on first examination of the

surface of a piece of charcoal, because it is often damaged and dirty. After a period of drying in an oven at 50°C, the charcoal will fracture readily. If care is taken to snap it along the radial longitudinal, tangential longitudinal and transverse planes, good surfaces for study can be produced. The specimens are mounted in plasticine on a microscope slide, and examined under the epi-illuminating microscope.

We have tried embedding and sectioning charcoal (with a diamond saw!) but so much material is lost in the process that it was found not to be worthwhile.

The very small fragments can be examined in the SEM, after coating, but generally the light microscope is adequate.

It can be decided if the makers of the fire had selected particular woods for their burning properties, or if the remains merely represent what was growing locally and easily accessible. An idea may be gathered about the particular vegetation of an area at particular times.

Some sites are very rich in waterlogged or dry, preserved wooden objects. The Sutton Hoo burial ship, for example, contained many wooden grave goods. Interesting examples from this site are some small pots with silver gilt rims. On excavation these were thought to be made from small gourds, fruit from the Cucurbitaceae. Microscopial study of thin sections showed the structure to be of walnut wood, probably from near the rootstock, where burr-wood could be obtained.

With improved techniques for recovering wooden wrecks and subsequently conserving them by special impregnation techniques, interest has increased in naval architecture. The timbers of a warship from the Punic wars were remarkably well preserved and were readily identified after many centuries in sea water.

An oak, Iron Age boat from Brigg in South Humberside also proved to be fascinating. No 'nails' were used to secure one timber to another, but the main logs were sewn together with twisted willow twigs passed through regularly pierced holes along the edges of the baulks of timbers.

In the bronze age, trackways were built across swampy ground in Somerset. The hazel faggots (*Corylus*) used in these were well preserved in the waterlogged conditions.

At the Jodrell Laboratory we look at archaeological material from all sorts of wooden objects; spear shafts, shields, buckets, right through to structural timbers. Much of this work is very time-consuming. Often some details of the anatomy are lost, and very careful comparisons with reference materials need to be made before identifications are given.

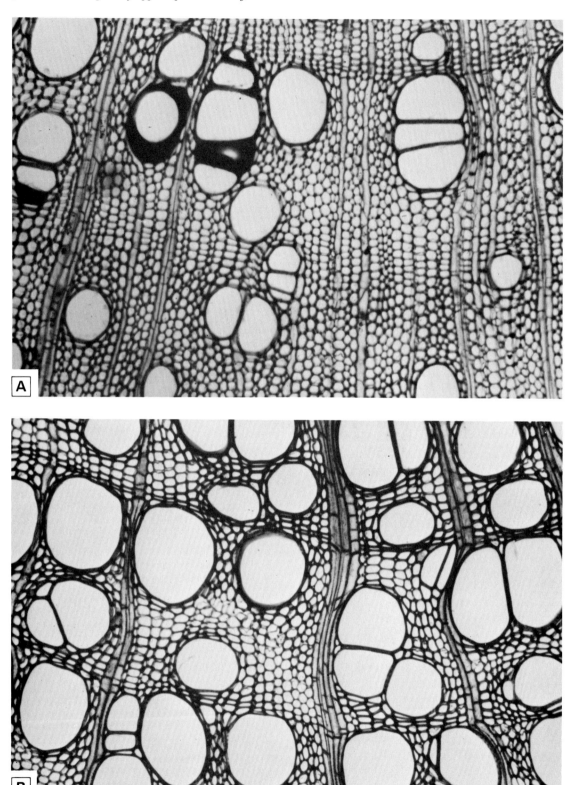

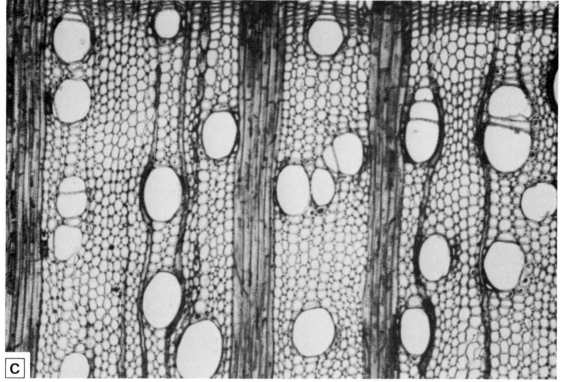

C

Fig. 9.5 *Acer pseudoplatanus* A, B, roots, grown under different conditions, T.S. A, from normal and B from waterlogged soils. C, normal trunk wood × 130.

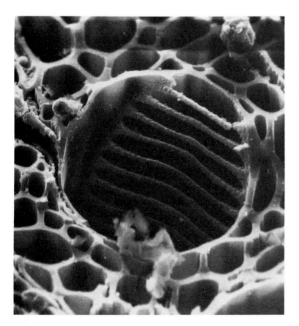

Fig. 9.6 Charcoal of *Alnus glutinosa* from a Romano–British site in London. Details of structure are well preserved, particularly the salariform perforation plate.

Because of the potentially enormous quantity of fragments of wood that could come from even one fire, we have to limit to a small number the sample of important specimens that we will accept from enquiries for each site.

Wood products

Archaeological plant remains other than from wood can sometimes be remarkably well preserved. The sandal shown in Fig. 9.7 from ancient Egypt is such an example. Papyrus is a major constituent of the sandal, and some *Borassus* palm is also present.

However, some of the samples are waterlogged and compressed. It is often possible to 'revive' such material. The secret is to section it in the compressed form and revive the sections, by floating them briefly in sodium hypochlorite solution or in chlor-zinc-iodine. Temporary mounts are best made in 50 per cent glycerine.

The structural properties of wood are utilized in modern building methods by using not only solid timber, but laminates, plywoods, chipboards, hardboards and the like. These materials are tested

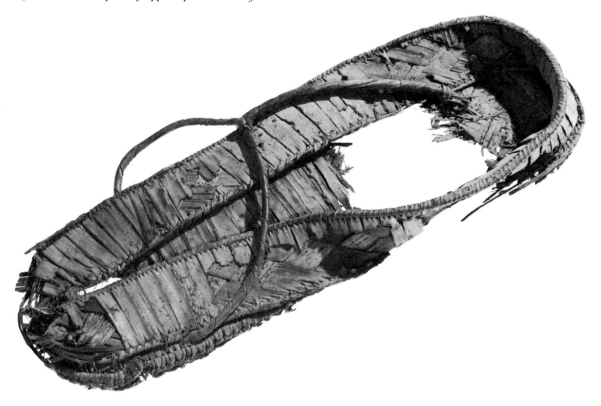

Fig. 9.7   An Egyptian sandal from antiquity, found to be made from papyrus (*Cyperus papyrus*) and *Borassus* species, palm.

to destruction so that their properties can be properly evaluated. Microscopic examination of the failure areas can give a good guide to areas of weakness.

Forensic applications

Forensic work often involves the identification of small pieces of plant material other than wood, although in addition to safe ballast, wood splinters might come from such things as windows, doors and their frames, weapons and the like, and thus play an important part in police work. A wide range of particles of plants may become attached to clothing or footwear which relate to the scene of a crime. If found on a suspect they may link him with the location of the crime. Clothing itself is made from a variety of fibres, a number of which may be of plant origin. Microscope slides of macerated textile fibres make up part of the reference collection of forensic laboratories.

Drug plants, such as *Cannabis sativa* frequently have diagnostic characters whereby quite small pieces may be identified microscopically.

An increasing number of plant species are being sold for consumption as drugs – some as adulterants, others as substitutes. It is a hard task to keep up with the introduction of additional species, particularly since the product is often in a very finely powdered form. Quite a lot of time and effort has to go into analysing such finely powdered drugs.

Anatomical characters can be used with such confidence for identification that they may contribute part of the evidence given under oath in court.

Postscript

In these pages we have seen the basic cellular structure of the vegetative parts of a number of common and less common plants described in simple terms. The evidence that plant anatomy is not just an academic subject has been drawn from a wide range of applications, many of economic importance, others of legal consequence and a number which simply served to answer intriguing questions.

Further Reading

Parry, J. W., 1962. *Spices. Their Morphology, Histology and Chemistry*, Chemical Publishing Company, New York.

Seiderman, J., 1966. *Stärke Atlas*, Paul Pary, Berlin.

Jackson, B. P. & Snowdon, D. W., 1968. *Powdered Vegetable Drugs*, J. & A. Churchill, London.

Trease, G. E. & Evans, W. C., 1972. *Pharmacognosy*, 10th edn, Baillière Tindall, London.

Wallace, T. E., 1967. *Textbook of Pharmacognosy*, 5th edn, J. & A. Churchill, London.

The following books may be difficult to find, but are worth the effort:

Hayward, H. E., 1938. *The Structure of Economic Plants*, Macmillan, New York.

Winton, A. L., Moeller, J. & Winton, K. B., 1916. *The Microscopy of Vegetable Foods*, John Wiley, New York.

Index

See also the illustrated glossary, Chapter 3, pp. 15–29. There are lists of families where particular characters may be seen on pp. 54 and 55. Some leaf and stem characters to be found in common plants from various parts of the world are on pages 55–7.

Some softwoods in which particular features can be found and some characters in the secondary xylem of selected hardwoods can be found at the end of Chapter 6, p. 75 and following.

Italic numerals indicate an illustration.